Gerzson Kéri

Compressed Chebyshev Polynomials and Multiple-Angle Formulas

II

Gerzson Kéri

Compressed Chebyshev Polynomials and Multiple-Angle Formulas II

GlobeEdit

Imprint

Cover image: www.ingimage.com

Publisher:
GlobeEdit
is a trademark of
Dodo Books Indian Ocean Ltd., member of the OmniScriptum S.R.L Publishing group
str. A.Russo 15, of. 61, Chisinau-2068, Republic of Moldova Europe
Printed at: see last page
ISBN: 978-620-0-63046-9

Compressed Chebyshev Polynomials and Multiple-Angle Formulas II

Kéri Gerzson

Contents

List of Tables

List of Figures

Introduction

The material of this publication can be regarded as the continuation of the author's previous publication [4] with the same title. It can also be regarded as a supplement, containing additional tables, longer listings for the tables that are given in [4], and additional figures.

The arrangement of the tables and figures is as follows:

- The first 10 tables summarize the most important general properties and identities with the four kinds of Chebyshev polynomials and their modified variants. These are put into words and statements (in the language of mathematical expressions) that concern arbitrary positive integer n.
- The next 26 tables are listings for a given interval of particular values of n, namely for $1 \le n \le 20$. Tables 11-14 simply list the compressed (modified) Chebyshev polynomials, while Tables 15-36 specify the irreducible factorization of not only the compressed Chebyshev polynomials, but also some related polynomials, where $1 \le n \le 20$ again.
- The factorizations shown in Tables 15-28 use the cyclotomic pre-polynomials $\Psi_n(x)$ from a non-contiguous domain that ends at the magnitude of 6 times the greatest degree (actually 20) of the polynomials to be factored in these tables. ($\Psi_{120}(x)$ occurs in Table 18 between the factors of $T_n^*(x) - 1$ when $n = 20$.) Therefore a much longer table that covers more than 10 pages for the parallel listing of cyclotomic pre-polynomials $\Psi_n(x)$ and cyclotomic polynomials $\Phi_n(x)$ for the interval $1 \le n \le 120$ is constructed as Table 37. Table 38 is a much shorter listing of $\Psi_n^{**}(x)$ that relates to the factors occuring in Tables 29-36.
- Eight figures are placed at the end in order to illustrate the connection between the roots of some compressed Chebyshev polynomial variants and the roots of their transforms.

Acknowledgment. The author would like to thank Lajos Szilassi for the very careful drawing of the figures.

Explanation, remarks and notices to the tables and figures

Explanation to Tables 1-3. $T_n(x)$, $U_n(x)$, $V_n(x)$, and $W_n(x)$, respectively, are the Chebyshev polynomials of the first, second, third, and fourth kind, respectively, as they are defined in [5].

$T_n^*(x)$, $U_n^*(x)$, $V_n^*(x)$ and $W_n^*(x)$ are the compressed (or in other words: modified) Chebyshev polynomials that are obtained from the original ones by simple linear transformations in the following way:

$$T_n^*(x) = 2T_n\left(\frac{x}{2}\right), U_n^*(x) = U_n\left(\frac{x}{2}\right), V_n^*(x) = V_n\left(\frac{x}{2}\right), W_n^*(x) = W_n\left(\frac{x}{2}\right).$$

$T_n^{**}(x)$ and $U_n^{**}(x)$ are formed from $T_n^*(x)$ and $U_n^*(x)$ by changing all negative coefficients to their absolute values, or by applying the transformations

$$T_n^{**}(x) = (-i)^n T_n^*(ix), \quad U_n^{**}(x) = (-i)^n U_n^*(ix).$$

In [4] it has been shown that the compressed Chebyshev polynomials have several advantages to the original ones. It is a further advantage, which has not been emphasized there, that, e.g., the factorization of $T_n^*(x)$ and the factorization of the polynomial $x^{2n}+1$ are fully synchron with each other, as all factors of the former are cyclotomic pre-polynomials, while the factors of the latter are the corresponding cyclotomic polynomials. Similar assertions are valid for the other three compressed Chebyshev polynomials $U_n^*(x)$, $V_n^*(x)$, and $W_n^*(x)$, respectively, if they are matched to the polynomials $\sum_{k=0}^{n} x^{2k}$, $\sum_{k=0}^{2n}(-1)^k x^k$, and $\sum_{k=0}^{2n} x^k$, respectively. (See the factorizational formulas of Tables 6–9.)

Remarks to Table 4. From the first identity of Table 3 it follows that the roots of $T_n^*(x) - h$ are

$$2\cos\theta_k,\ (k = 0, 1, \ldots, n-1),\ \text{where } \theta_k = \frac{\arccos(h/2) + 2k\pi}{n} \tag{1}$$

for any $|h| \le 2$. By specifying (1) to $h = 2, -2, 0, 1, -1$, the middle column of Table 4 can be built. The rightmost column of Table 4, where the roots are arranged in monotonic decreasing order, contain mainly double roots for $T_n^*(x) \mp 2$. The remaining part of the table shows only singular roots.

One can observe that for the values of h under consideration, both the roots are rational and the arguments of the cosine function contained in the table are always rational times of π (i.e. these arguments themselves are rational in degrees). The converse of this assertion is also true in the following sense: If $|h| \le 2$ and $h = T_n^*(2\cos\theta) = 2\cos n\theta$ where h is rational and also θ is rational in degrees, then h cannot be other than $h = 0, \pm 1, \pm 2$. This follows from a more general theorem of Niven [6, Corollary 3.12] which we quote here literally: "*If* θ *is rational in degrees, say* $\theta = 2\pi r$ *for some rational number* r*, then the only rational values of the trigonometric functions of* θ *are as follows:* $\sin\theta, \cos\theta = 0, \pm\frac{1}{2}, \pm 1; \sec\theta, \csc\theta = \pm 1, \pm 2; \tan\theta, \cot\theta = 0, \pm 1.$"

Remark to Table 5. As regards the roots of $U_n^*(x) - h$ we cannot serve similar convincing explanation that we did for the case of $T_n^*(x) - h$. It appered, however, that only the cases $h = 0$ and ± 1 can produce interesting results from the point of you of roots and factorization.

Remarks to Tables 6-10. As regards Tables 6, 7, and 10 we refer to [4, Chapter 4]. The contents of Tables 8 and 9 can be proved by using similar methods for the modified Chebyshev polynomials of the third and the fourth kinds.

For the sake of readers who are unable to look into [4], next we provide a short explanation regarding to the *overline* transformation and the cyclotomic polynomials.

Let us consider the transformation of an arbitrary polynomial $P_n(x)$ of degree n into the polynomial $\overline{P_n(x)} = x^n P_n\left(x + \frac{1}{x}\right)$. It can be seen easily, that the *overline* transformation is multiplicative, i.e. $\overline{P_n(x)Q_m(x)} = \overline{P_n(x)} \cdot \overline{Q_m(x)}$. Consequently, if the polynomial $P_n(x)$ can be divided by the polynomial $Q_m(x)$, then $\overline{P_n(x)/Q_m(x)} = \overline{P_n(x)}/\overline{Q_m(x)}$.

Two more useful rules of thumb regarding the *overline* transformation are as follows: For arbitrary real numbers a, b, except $a = 0$, we have $\overline{aP_n(x) + b} = a\overline{P_n(x)} + bx^n$. For any two polynomials $P_n(x)$ and $Q_m(x)$ where for their degrees $n \geq m$ holds, we have $\overline{P_n(x) + Q_m(x)} = \overline{P_n(x)} + x^{n-m}\overline{Q_m(x)}$.

"The nth cyclotomic polynomial, for any positive integer n, is the unique irreducible polynomial with integer coefficients that is a divisor of $x^n - 1$ and is not a divisor of $x^k - 1$ for any $k < n$. Its roots are all nth primitive roots of unity $e^{2i\pi\frac{k}{n}}$, where k runs over the positive integers not greater than n and coprime to n (and i is the imaginary unit)." (The sentences between quotation marks are quoted literally from [9].) In other words, the nth cyclotomic polynomial is equal to

$$\Phi_n(x) = \prod_{k=1}^{\varphi(n)}(x - \xi_k) = \prod_{\substack{1 \leq k \leq n \\ \gcd(k,n)=1}} (x - e^{\frac{2k\pi i}{n}})$$

where ξ_k, $(k = 1, 2, \varphi(n))$ is the nth primitive root of unity and $\varphi(n)$ is the Euler's totient function. As for $n > 1$, n is not coprime to itself, therefore

$$\Phi_n(x) = \prod_{\substack{1 \le k < n \\ \gcd(k,n)=1}} (x - e^{\frac{2k\pi i}{n}})$$

is also a correct formula, except for $n = 1$. The cyclotomic polynomials have the following properties:

The cyclotomic polynomials are monic polynomials with integer coefficients that are irreducible over the field of the rational numbers. Except for n equal to 1 or 2, they are palindromics of even degree. Furthermore, each nth root of unity is a primitive dth root of unity for a unique d dividing n, i.e.

$$x^n - 1 = \prod_{d|n} \Phi_d(x).$$

If n is an odd integer greater than 1, then

$$\Phi_{2n}(x) = \Phi_n(-x).$$

For arbitrary n greater than 2 there exists a polynomial $\Psi_n(x)$ of degree $\varphi(n)/2$, for which $\overline{\Psi_n(x)} = \Phi_n(x)$, that may be called *cyclotomic pre-polynomial*, and for which

$$\Psi_n(x) = \prod_{\substack{0 < k < n/2 \\ \gcd(k,n)=1}} \left(x - 2\cos\frac{2k\pi}{n}\right).$$

As for any even integer n that is greater than 2, $n/2$ is not coprime to n, therefore

$$\Psi_n(x) = \prod_{\substack{0 < k \le n/2 \\ \gcd(k,n)=1}} \left(x - 2\cos\frac{2k\pi}{n}\right).$$

is also a correct formula for any n, for which it is applicable.

Similarly to the cyclotomic polynomials, also the cyclotomic pre-polynomials $\Psi_n(x)$ have only integer coefficients and are irreducible over the field of the rational numbers, see [2, Theorem 4.1].

We shall use $\Psi_n(x)$ also for $n = 1$ and 2, that we define as follows:

$$\Psi_1(x) = \sqrt{x - 2}, \quad \Psi_2(x) = \sqrt{x + 2}.$$

These are not polynomials, but for their squares (that are polynomials) we have

$$\overline{[\Psi_1(x)]^2} = [\Phi_1(x)]^2 \text{ and } \overline{[\Psi_2(x)]^2} = [\Phi_2(x)]^2.$$

As regards Tables 11-36, it is unnecessary to add any additional remarks to the description given in the Introduction.

Remarks to Table 37. Many cases have a prompt way of handling. They are as follows:

a) If p is an odd prime, then

$$\Psi_p(x) = W^*_{(p-1)/2}(x) \text{ and } \Phi_p(x) = \overline{W^*_{(p-1)/2}(x)} = \sum_{k=0}^{p-1} x^k.$$

Justification: Consider the formula for the factorization of $W^*_n(x)$ in the second line of Table 9. There it can be seen that $W^*_n(x)$ is irreducible and equals to $\Psi_{2n+1}(x)$ whenever $2n+1$ is prime.

b) Under the same condition

$$\Psi_{2p}(x) = V^*_{(p-1)/2}(x) \text{ and } \Phi_{2p}(x) = \overline{V^*_{(p-1)/2}(x)} = \sum_{k=0}^{p-1} (-1)^k x^k.$$

This can be justified similarly to the previous case by considering the first line of Table 9.

c) If $n = 2^q$ where $q \geq 2$, then

$$\Psi_n(x) = T^*_{n/4}(x) \text{ and } \Phi_n(x) = \overline{T^*_{n/4}(x)} = x^{n/2} + 1.$$

(Consider the first line of Table 7).

d) If $n = 3^r$ where $r \geq 1$, then

$$\Psi_n(x) = T^*_{n/3}(x) + 1 \text{ and } \Phi_n(x) = \overline{T^*_{n/3}(x) + 1} = \sum_{k=0}^{2} x^{kn/3}.$$

(Consider the fifth line of Table 7).

e) If $n = 2^q 3^r$ where $q, r \geq 1$, then

$$\Psi_n(x) = T^*_{n/6}(x) - 1 \text{ and } \Phi_n(x) = \overline{T^*_{n/6}(x) - 1} = \sum_{k=0}^{2} (-1)^k x^{kn/6}.$$

(Consider the fourth line of Table 7).

The second parts of assertions a)-e) that applies to $\Phi_n(x)$ are well known from the theory of cyclotomic polynomials, see e.g. [9]. Two more cases: f) and g) below regarding only to $\Phi_n(x)$ can also be found in [9].

f) If $n = p^r$ where p is a prime, greater than 3, and $r \geq 2$, then

$$\Phi_n(x) = \sum_{k=0}^{p-1} x^{kn/p}.$$

g) If $n = 2^q p^r$ where p is a prime, greater than 3, and $q, r \geq 1$, then

$$\Phi_n(x) = \sum_{k=0}^{p-1} (-1)^k x^{kn/p/2}.$$

To generate the cyclotomic pre-polynomials $\Psi_n(x)$ for those values of n that do not meet either of the conditions enumerated above as a)-e), we can use the same formulas for the factorization of $W_n^*(x)$ whenever $n \equiv 1 \pmod 2$, $V_n^*(x)$ whenever $n \equiv 2 \pmod 4$, and $T_n^*(x)$ whenever $n \equiv 0 \pmod 4$. The determination of $\Psi_n(x)$ does not remain as straightforward as for the cases enumerated above, because some computational works are required for the cases not yet arranged so far. For this computation, several applications are available from the web, such as [13] and [14].

Also [13], [14] or similar applications can be applied for computing the polynomials $\Psi_n^{**}(x)$ for odd n, by the factorization of $U_n^{**}(x)$ for a set of even values of n. (Cf. Tables 10 and 34.)

Remarks to Figures 1–8. Let $P_n(x)$ be a polynomial of degree n, such that it has n real roots: $x_1, x_2, \ldots, x_n$, each lying in the interval $-2 \leq x_k \leq +2$. (All kinds of the compressed Chebyshev polynomials as well as their translations that occur in Tables 7 and 9 meet this condition.) Then x_k can be written as $x_k = 2\cos\theta_k$ that gives

$$P_n(x) = \prod_{k=1}^{n} (x - 2\cos\theta_k).$$

Consequently

$$\overline{P_n(x)} = \prod_{k=1}^{n} (x^2 - 2x\cos\theta_k + 1) = \prod_{k=1}^{n} (x - e^{i\theta_k})(x - e^{-i\theta_k}).$$

This means that to every root of $P_n(x)$ there corresponds two conjugate complex roots of $\overline{P_n(x)}$, which may coincide and provide a real root with multiplicity 2 when $|x_k|$=2.

Figures 1–8 illustrate the connection between the roots of $P_n(x)$ and $\overline{P_n(x)}$ when $n = 3$ or $n = 4$, and the role of $P_n(x)$ is played by $T_n^*(x)$ or $U_n^*(x)$, or some of their translations.

Table 1: Explicit expressions for the original and for the compressed Chebyshev polynomials

$$T_n(x) = \sum_{k=0}^{\lfloor n/2 \rfloor} (-1)^k \left(\!\!\binom{n-k}{k}\!\!\right) 2^{n-2k-1} x^{n-2k}$$

$$U_n(x) = \sum_{k=0}^{\lfloor n/2 \rfloor} (-1)^k \binom{n-k}{k} 2^{n-2k} x^{n-2k}$$

$$V_n(x) = \sum_{k=0}^{\lfloor n/2 \rfloor} (-1)^k \left[\binom{n-k}{k} 2^{n-2k} x^{n-2k} - \binom{n-1-k}{k} 2^{n-1-2k} x^{n-1-2k} \right] + \varepsilon$$

or otherwise: $V_n(x) = \sum_{k=0}^{n} (-1)^{\lceil k/2 \rceil} \binom{\lfloor n-k/2 \rfloor}{\lfloor k/2 \rfloor} 2^{n-k} x^{n-k}$

$$W_n(x) = \sum_{k=0}^{\lfloor n/2 \rfloor} (-1)^k \left[\binom{n-k}{k} 2^{n-2k} x^{n-2k} + \binom{n-1-k}{k} 2^{n-1-2k} x^{n-1-2k} \right] + \varepsilon$$

or otherwise: $W_n(x) = \sum_{k=0}^{n} (-1)^{\lfloor k/2 \rfloor} \binom{\lfloor n-k/2 \rfloor}{\lfloor k/2 \rfloor} 2^{n-k} x^{n-k}$

$$T_n^*(x) = \sum_{k=0}^{\lfloor n/2 \rfloor} (-1)^k \left(\!\!\binom{n-k}{k}\!\!\right) x^{n-2k}$$

$$U_n^*(x) = \sum_{k=0}^{\lfloor n/2 \rfloor} (-1)^k \binom{n-k}{k} x^{n-2k}$$

$$V_n^*(x) = \sum_{k=0}^{\lfloor n/2 \rfloor} (-1)^k \left[\binom{n-k}{k} x^{n-2k} - \binom{n-1-k}{k} x^{n-1-2k} \right] + \varepsilon$$

or otherwise: $V_n^*(x) = \sum_{k=0}^{n} (-1)^{\lceil k/2 \rceil} \binom{\lfloor n-k/2 \rfloor}{\lfloor k/2 \rfloor} x^{n-k}$

$$W_n^*(x) = \sum_{k=0}^{\lfloor n/2 \rfloor} (-1)^k \left[\binom{n-k}{k} x^{n-2k} + \binom{n-1-k}{k} x^{n-1-2k} \right] + \varepsilon$$

or otherwise: $W_n^*(x) = \sum_{k=0}^{n} (-1)^{\lfloor k/2 \rfloor} \binom{\lfloor n-k/2 \rfloor}{\lfloor k/2 \rfloor} x^{n-k}$

$$T_n^{**}(x) = \sum_{k=0}^{\lfloor n/2 \rfloor} \left(\!\!\binom{n-k}{k}\!\!\right) x^{n-2k}$$

$$U_n^{**}(x) = \sum_{k=0}^{\lfloor n/2 \rfloor} \binom{n-k}{k} x^{n-2k}$$

where

$$\left(\!\!\binom{n}{k}\!\!\right) = \begin{cases} \binom{n}{k} & \text{if} \quad k = 0, \\ \binom{n}{k} + \binom{n-1}{k-1} & \text{if} \quad 1 \le k \le n \end{cases}$$

and

$$\varepsilon = \begin{cases} 0 & \text{if} \quad n \equiv 1 \pmod 2 \\ 1 & \text{if} \quad n \equiv 0 \pmod 4 \\ -1 & \text{if} \quad n \equiv 2 \pmod 4 \end{cases}$$

Table 2: Recursive rule for the different kinds of Chebyshev polynomials

$T_{n+1}(x) = 2xT_n(x) - T_{n-1}(x)$	$T^*_{n+1}(x) = xT^*_n(x) - T^*_{n-1}(x)$
$U_{n+1}(x) = 2xU_n(x) - U_{n-1}(x)$	$U^*_{n+1}(x) = xU^*_n(x) - U^*_{n-1}(x$
$V_{n+1}(x) = 2xV^*_n(x) - V_{n-1}(x)$	$V^*_{n+1}(x) = xV^*_n(x) - V^*_{n-1}(x)$
$W_{n+1}(x) = 2xW^*_n(x) - W_{n-1}(x)$	$W^*_{n+1}(x) = xW^*_n(x) - W^*_{n-1}(x)$
$T^{**}_{n+1}(x) = xT^{**}_n(x) + T^{**}_{n-1}(x)$	$U^{**}_{n+1}(x) = xU^{**}_n(x) + U^{**}_{n-1}(x)$

Table 3: Multiple angles expressed by the modified Chebyshev polynomials

	$2\cos n\theta$	$=$	$T^*_n(2\cos\theta)$
	$2\cosh n\theta$	$=$	$T^*_n(2\cosh\theta)$
n odd: n even:	$2\sin n\theta$ $2\cos n\theta$	$\}=$	$(-1)^{\lfloor n/2\rfloor}T^*_n(2\sin\theta)$
n odd: n even:	$2\sinh n\theta$ $2\cosh n\theta$	$\}=$	$T^{**}_n(2\sinh\theta)$
	$\frac{\sin(n+1)\theta}{\sin\theta}$	$=$	$U^*_n(2\cos\theta)$
	$\frac{\sinh(n+1)\theta}{\sinh\theta}$	$=$	$U^*_n(2\cosh\theta)$
n odd: n even:	$\frac{\sin(n+1)\theta}{\cos\theta}$ $\frac{\cos(n+1)\theta}{\cos\theta}$	$\}=$	$(-1)^{\lfloor n/2\rfloor}U^*_n(2\sin\theta)$
n odd: n even:	$\frac{\sinh(n+1)\theta}{\cosh\theta}$ $\frac{\cosh(n+1)\theta}{\cosh\theta}$	$\}=$	$U^{**}_n(2\sinh\theta)$
	$\cos\left(n+\frac{1}{2}\right)\theta$	$=$	$\left(\cos\frac{1}{2}\theta\right)V^*_n(2\cos\theta)$
	$\sin\left(n+\frac{1}{2}\right)\theta$	$=$	$\left(\sin\frac{1}{2}\theta\right)W^*_n(2\cos\theta)$

Table 4: Roots of $T_n^*(x) - h \;\; (h = \pm 2, 0, \pm 1)$

	roots given according to (1)	roots in monotonic decreasing order
$T_n^*(x) - 2$	$2\cos\frac{2k\pi}{n}$, $(k = 0, 1, \ldots, n-1)$	$2\cos\frac{2k\pi}{n}$, $(k = 0, 1, \ldots, \lfloor\frac{n}{2}\rfloor)$
$T_n^*(x) + 2$	$2\cos\frac{(2k+1)\pi}{n}$, $(k = 0, 1, \ldots, n-1)$	$2\cos\frac{(2k+1)\pi}{n}$, $(k = 0, 1, \ldots, \lfloor\frac{n-1}{2}\rfloor)$
$T_n^*(x)$	$2\cos\frac{(2k+1/2)\pi}{n}$, $(k = 0, 1, \ldots, n-1)$	$2\cos\frac{(2k+1)\pi}{2n}$, $(k = 0, 1, \ldots, n-1)$
$T_n^*(x) - 1$	$2\cos\frac{(2k+1/3)\pi}{n}$, $(k = 0, 1, \ldots, n-1)$	$2\cos\frac{(2k+1)\pi}{3n}$, $(k = 0, 2, 3, 5 \ldots, \lfloor\frac{3n-2}{2}\rfloor)$
$T_n^*(x) + 1$	$2\cos\frac{(2k+2/3)\pi}{n}$, $(k = 0, 1, \ldots, n-1)$	$2\cos\frac{(2k+2)\pi}{3n}$, $(k = 0, 1, 3, 4, \ldots, \lfloor\frac{3n-3}{2}\rfloor)$

Table 5: Roots of $U_n^*(x) - h \;\; (h = 0, \pm 1)$

$U_n^*(x)$	$2\cos\frac{k\pi}{n+1}$, $(k = 1, \ldots, n)$
$U_n^*(x) - 1$	$2\cos\frac{2k\pi}{n}$, $(k = 1, 2, \ldots, \lfloor\frac{n-1}{2}\rfloor)$; $2\cos\frac{(2k+1)\pi}{n+2}$, $(k = 0, 1, \ldots, \lfloor\frac{n}{2}\rfloor)$
$U_n^*(x) + 1$	$2\cos\frac{(2k+1)\pi}{n}$, $(k = 0, 1, \ldots, \lfloor\frac{n-2}{2}\rfloor)$; $2\cos\frac{2k\pi}{n+2}$, $(k = 1, 2, \ldots, \lfloor\frac{n+1}{2}\rfloor)$

Table 6: Irreducible factorization of $\overline{T_n^*(x) - h}$ $(h = 0, \pm 2, \pm 1)$ and $\overline{U_n^*(x) - h}$ $(h = 0, \pm 1)$

polynomial		factorization	number of factors	OEIS
$\overline{T_n^*(x)}$	$= x^{2n} + 1$	$\prod_{\substack{d\mid 4n \\ d\nmid 2n}} \Phi_d(x)$	$\tau(2n) - \tau(n)$	
$\overline{T_n^*(x) - 2}$	$= (x^n - 1)^2$	$\prod_{d\mid n} [\Phi_d(x)]^2$	$2\tau(n)$	A062011
$\overline{T_n^*(x) + 2}$	$= (x^n + 1)^2$	$\prod_{\substack{d\mid 2n \\ d\nmid n}} [\Phi_d(x)]^2$	$2\tau(2n) - 2\tau(n)$	A054844
$\overline{T_n^*(x) - 1}$	$= \frac{x^{3n}+1}{x^n+1}$	$\prod_{\substack{d\mid 6n \\ d\nmid 2n \\ d\nmid 3n}} \Phi_d(x)$	$\tau(6n) - \tau(2n) - \tau(3n) + \tau(n)$	
$\overline{T_n^*(x) + 1}$	$= \frac{x^{3n}-1}{x^n-1}$	$\prod_{\substack{d\mid 3n \\ d\nmid n}} \Phi_d(x)$	$\tau(3n) - \tau(n)$	
$\overline{U_n^*(x)}$	$= \frac{x^{2n+2}-1}{x^2-1}$	$\prod_{\substack{d\geq 3 \\ d\mid 2n+2}} \Phi_d(x)$	$\tau(2n+2) - 2$	
$\overline{U_n^*(x) - 1}$	$= \frac{(x^{2n+4}-1)(x^n-1)}{(x^{n+2}-1)(x^2-1)}$	$\prod_{\substack{d\geq 3 \\ d\mid n}} \Phi_d(x) \prod_{\substack{d\geq 3 \\ d\mid 2n+4 \\ d\nmid n+2}} \Phi_d(x)$	$\tau(2n+4) + \tau(n) - \tau(n+2) - 2$	
$\overline{U_n^*(x) + 1}$	$= \frac{(x^{2n}-1)(x^{n+2}-1)}{(x^n-1)(x^2-1)}$	$\prod_{\substack{d\geq 3 \\ d\mid n+2}} \Phi_d(x) \prod_{\substack{d\geq 3 \\ d\mid 2n \\ d\nmid n}} \Phi_d(x)$	$\tau(2n) + \tau(n+2) - \tau(n) - 2$	

Table 7: Irreducible factorization of $T_n^*(x) - h$ $(h = 0, \pm 2, \pm 1)$ and $U_n^*(x) - h$ $(h = 0, \pm 1)$

polynomial	factorization	number of factors	OEIS
$T_n^*(x)$	$\prod_{\substack{d \mid 4n \\ d \nmid 2n}} \Psi_d(x)$	$\tau(2n) - \tau(n)$	A001227
$T_n^*(x) - 2$	$\prod_{d \mid n} [\Psi_d(x)]^2$	$2\tau(n) - 1$ $(n$ odd$)$ $2\tau(n) - 2$ $(n$ even$)$	A086369
$T_n^*(x) + 2$	$\prod_{\substack{d \mid 2n \\ d \nmid n}} [\Psi_d(x)]^2$	$2\tau(2n) - 2\tau(n) - 1$ $(n$ odd$)$ $2\tau(2n) - 2\tau(n)$ $(n$ even$)$	A086374
$T_n^*(x) - 1$	$\prod_{\substack{d \mid 6n \\ d \nmid 2n \\ d \nmid 3n}} \Psi_d(x)$	$\tau(6n) - \tau(2n) - \tau(3n) + \tau(n)$	--
$T_n^*(x) + 1$	$\prod_{\substack{d \mid 3n \\ d \nmid n}} \Psi_d(x)$	$\tau(3n) - \tau(n)$	--
$U_n^*(x)$	$\prod_{\substack{d \geq 3 \\ d \mid 2n+2}} \Psi_d(x)$	$\tau(2n+2) - 2$	A086327
$U_n^*(x) - 1$	$\prod_{\substack{d \geq 3 \\ d \mid n}} \Psi_d(x) \prod_{\substack{d \geq 3 \\ d \mid 2n+4 \\ d \nmid n+2}} \Psi_d(x)$	$\tau(2n+4) + \tau(n) - \tau(n+2) - 2$	A086389
$U_n^*(x) + 1$	$\prod_{\substack{d \geq 3 \\ d \mid n+2}} \Psi_d(x) \prod_{\substack{d \geq 3 \\ d \mid 2n \\ d \nmid n}} \Psi_d(x)$	$\tau(2n) + \tau(n+2) - \tau(n) - 2$	A086375

Table 8: Irreducible factorization of $\overline{V_n^*(x)-h}$ $(h=0,\pm1)$ and $\overline{W_n^*(x)-h}$ $(h=0,\pm1)$

polynomial	factorization	number of factors
$\overline{V_n^*(x)} = \frac{x^{2n+1}+1}{x+1}$	$\prod_{\substack{d\geq3\\ d\mid 4n+2\\ d\nmid 2n+1}} \Phi_d(x)$	$\tau(2n+1)-1$
$\overline{W_n^*(x)} = \frac{x^{2n+1}-1}{x-1}$	$\prod_{\substack{d\geq3\\ d\mid 2n+1}} \Phi_d(x)$	
$\overline{V_n^*(x)-1} = \frac{(x^n-1)(x^{n+1}-1)}{x+1}$	$\prod_{\substack{d\neq2\\ d\mid n}} \Phi_d(x) \prod_{\substack{d\neq2\\ d\mid n+1}} \Phi_d(x)$	$\tau(n+1)+\tau(n)-1$
$\overline{V_n^*(x)+1} = \frac{(x^n+1)(x^{n+1}+1)}{x+1}$	$\prod_{\substack{d\neq2\\ d\mid 2n\\ d\nmid n}} \Phi_d(x) \prod_{\substack{d\neq2\\ d\mid 2n+2\\ d\nmid n+1}} \Phi_d(x)$	$\tau(2n+2)-\tau(n+1)+\tau(2n)-\tau(n)-1$
$\overline{W_n^*(x)-1} = \frac{(x^n-1)(x^{n+1}+1)}{x-1}$	$\prod_{\substack{d\geq2\\ d\mid n}} \Phi_d(x) \prod_{\substack{d\mid 2n+2\\ d\nmid n+1}} \Phi_d(x)$	$\tau(2n+2)-\tau(n+1)+\tau(n)-1$
$\overline{W_n^*(x)+1} = \frac{(x^n+1)(x^{n+1}-1)}{x-1}$	$\prod_{\substack{d\geq2\\ d\mid n+1}} \Phi_d(x) \prod_{\substack{d\mid 2n\\ d\nmid n}} \Phi_d(x)$	$\tau(2n)-\tau(n)+\tau(n+1)-1$

Table 9: Irreducible factorization of $V_n^*(x)-h$ $(h=0,\pm1)$ and $W_n^*(x)-h$ $(h=0,\pm1)$

polynomial	factorization	number of factors	OEIS
$V_n^*(x)$	$\prod_{\substack{d\geq3\\ d\mid 4n+2\\ d\nmid 2n+1}} \Psi_d(x)$	$\tau(2n+1)-1$	A095374
$W_n^*(x)$	$\prod_{\substack{d\geq3\\ d\mid 2n+1}} \Psi_d(x)$		
$V_n^*(x)-1$	$\prod_{\substack{d\neq2\\ d\mid n}} \Psi_d(x) \prod_{\substack{d\neq2\\ d\mid n+1}} \Psi_d(x)$	$\tau(n+1)+\tau(n)-2$	--
$V_n^*(x)+1$	$\prod_{\substack{d\neq2\\ d\mid 2n\\ d\nmid n}} \Psi_d(x) \prod_{\substack{d\neq2\\ d\mid 2n+2\\ d\nmid n+1}} \Psi_d(x)$	$\tau(2n+2)-\tau(n+1)+\tau(2n)-\tau(n)-1$	--
$W_n^*(x)-1$	$\prod_{\substack{d\geq2\\ d\mid n}} \Psi_d(x) \prod_{\substack{d\mid 2n+2\\ d\nmid n+1}} \Psi_d(x)$	$\tau(2n+2)-\tau(n+1)+\tau(n)-1$ (n odd) $\tau(2n+2)-\tau(n+1)+\tau(n)-2$ (n even)	--
$W_n^*(x)+1$	$\prod_{\substack{d\geq2\\ d\mid n+1}} \Psi_d(x) \prod_{\substack{d\mid 2n\\ d\nmid n}} \Psi_d(x)$	$\tau(2n)-\tau(n)+\tau(n+1)-2$ (n odd) $\tau(2n)-\tau(n)+\tau(n+1)-1$ (n even)	--

Table 10: Factorization of $T_n^{**}(x) - h \ \ (h = 0, \pm 2, \pm 1)$ and $U_n^{**}(x) - h \ \ (h = 0, \pm 1)$ (irreducibility questionable)

$$
\begin{aligned}
T_n^{**}(x) &= \prod_{\substack{d|4n \\ d\nmid 2n}} \Psi_d^{**}(x) \\
U_n^{**}(x) &= \prod_{\substack{d\geq 3 \\ d|2n+2 \\ d\not\equiv 2 \pmod 4}} \Psi_d^{**}(x) \\
T_{2n}^{**}(x) + 2\cdot(-1)^n &= \prod_{\substack{d|4n \\ d\nmid 2n}} [\Psi_d^{**}(x)]^2 \\
T_{2n}^{**}(x) - 2\cdot(-1)^n &= \prod_{\substack{d|2n \\ d\not\equiv 2 \pmod 4}} [\Psi_d^{**}(x)]^2 \\
T_{2n}^{**}(x) + (-1)^n &= \prod_{\substack{d|6n \\ d\nmid 2n \\ d\not\equiv 2 \pmod 4}} \Psi_d^{**}(x) \\
T_{2n}^{**}(x) - (-1)^n &= \prod_{\substack{d|12n \\ d\nmid 4n \\ d\nmid 6n}} \Psi_d^{**}(x) \\
U_{2n}^{**}(x) + (-1)^n &= \prod_{\substack{d|4n \\ d\nmid 2n}} \Psi_d^{**}(x) \prod_{\substack{d\geq 3 \\ d|2n+2 \\ d\not\equiv 2 \pmod 4}} \Psi_d^{**}(x) \\
U_{2n}^{**}(x) - (-1)^n &= \prod_{\substack{d\geq 3 \\ d|2n \\ d\not\equiv 2 \pmod 4}} \Psi_d^{**}(x) \prod_{\substack{d\geq 3 \\ d|4n+4 \\ d\nmid 2n+2}} \Psi_d^{**}(x)
\end{aligned}
$$

Table 11: List of modified Chebyshev polynomials of the first kind for $1 \leq n \leq 20$

n	$T_n^*(x)$
1	x
2	$x^2 - 2$
3	$x^3 - 3x$
4	$x^4 - 4x^2 + 2$
5	$x^5 - 5x^3 + 5x$
6	$x^6 - 6x^4 + 9x^2 - 2$
7	$x^7 - 7x^5 + 14x^3 - 7x$
8	$x^8 - 8x^6 + 20x^4 - 16x^2 + 2$
9	$x^9 - 9x^7 + 27x^5 - 30x^3 + 9x$
10	$x^{10} - 10x^8 + 35x^6 - 50x^4 + 25x^2 - 2$
11	$x^{11} - 11x^9 + 44x^7 - 77x^5 + 55x^3 - 11x$
12	$x^{12} - 12x^{10} + 54x^8 - 112x^6 + 105x^4 - 36x^2 + 2$
13	$x^{13} - 13x^{11} + 65x^9 - 156x^7 + 182x^5 - 91x^3 + 13x$
14	$x^{14} - 14x^{12} + 77x^{10} - 210x^8 + 294x^6 - 196x^4 + 49x^2 - 2$
15	$x^{15} - 15x^{13} + 90x^{11} - 275x^9 + 450x^7 - 378x^5 + 140x^3 - 15x$
16	$x^{16} - 16x^{14} + 104x^{12} - 352x^{10} + 660x^8 - 672x^6 + 336x^4 - 64x^2 + 2$
17	$x^{17} - 17x^{15} + 119x^{13} - 442x^{11} + 935x^9 - 1122x^7 + 714x^5 - 204x^3 + 17x$
18	$x^{18} - 18x^{16} + 135x^{14} - 546x^{12} + 1287x^{10} - 1782x^8 + 1386x^6 - 540x^4 + 81x^2 - 2$
19	$x^{19} - 19x^{17} + 152x^{15} - 665x^{13} + 1729x^{11} - 2717x^9 + 2508x^7 - 1254x^5 + 285x^3 - 19x$
20	$x^{20} - 20x^{18} + 170x^{16} - 800x^{14} + 2275x^{12} - 4004x^{10} + 4290x^8 - 2640x^6 + 825x^4 - 100x^2 + 2$

Table 12: List of modified Chebyshev polynomials of the second kind for $1 \leq n \leq 20$

n	$U_n^*(x)$
1	x
2	$x^2 - 1$
3	$x^3 - 2x$
4	$x^4 - 3x^2 + 1$
5	$x^5 - 4x^3 + 3x$
6	$x^6 - 5x^4 + 6x^2 - 1$
7	$x^7 - 6x^5 + 10x^3 - 4x$
8	$x^8 - 7x^6 + 15x^4 - 10x^2 + 1$
9	$x^9 - 8x^7 + 21x^5 - 20x^3 + 5x$
10	$x^{10} - 9x^8 + 28x^6 - 35x^4 + 15x^2 - 1$
11	$x^{11} - 10x^9 + 36x^7 - 56x^5 + 35x^3 - 6x$
12	$x^{12} - 11x^{10} + 45x^8 - 84x^6 + 70x^4 - 21x^2 + 1$
13	$x^{13} - 12x^{11} + 55x^9 - 120x^7 + 126x^5 - 56x^3 + 7x$
14	$x^{14} - 13x^{12} + 66x^{10} - 165x^8 + 210x^6 - 126x^4 + 28x^2 - 1$
15	$x^{15} - 14x^{13} + 78x^{11} - 220x^9 + 330x^7 - 252x^5 + 84x^3 - 8x$
16	$x^{16} - 15x^{14} + 91x^{12} - 286x^{10} + 495x^8 - 462x^6 + 210x^4 - 36x^2 + 1$
17	$x^{17} - 16x^{15} + 105x^{13} - 364x^{11} + 715x^9 - 792x^7 + 462x^5 - 120x^3 + 9x$
18	$x^{18} - 17x^{16} + 120x^{14} - 455x^{12} + 1001x^{10} - 1287x^8 + 924x^6 - 330x^4 + 45x^2 - 1$
19	$x^{19} - 18x^{17} + 136x^{15} - 560x^{13} + 1365x^{11} - 2002x^9 + 1716x^7 - 792x^5 + 165x^3 - 10x$
20	$x^{20} - 19x^{18} + 153x^{16} - 680x^{14} + 1820x^{12} - 3003x^{10} + 3003x^8 - 1716x^6 + 495x^4 - 55x^2 + 1$

Table 13: List of modified Chebyshev polynomials of the third kind for $1 \leq n \leq 20$

n	$V_n^*(x)$
1	$x - 1$
2	$x^2 - x - 1$
3	$x^3 - x^2 - 2x + 1$
4	$x^4 - x^3 - 3x^2 + 2x + 1$
5	$x^5 - x^4 - 4x^3 + 3x^2 + 3x - 1$
6	$x^6 - x^5 - 5x^4 + 4x^3 + 6x^2 - 3x - 1$
7	$x^7 - x^6 - 6x^5 + 5x^4 + 10x^3 - 6x^2 - 4x + 1$
8	$x^8 - x^7 - 7x^6 + 6x^5 + 15x^4 - 10x^3 - 10x^2 + 4x + 1$
9	$x^9 - x^8 - 8x^7 + 7x^6 + 21x^5 - 15x^4 - 20x^3 + 10x^2 + 5x - 1$
10	$x^{10} - x^9 - 9x^8 + 8x^7 + 28x^6 - 21x^5 - 35x^4 + 20x^3 + 15x^2 - 5x - 1$
11	$x^{11} - x^{10} - 10x^9 + 9x^8 + 36x^7 - 28x^6 - 56x^5 + 35x^4 + 35x^3 - 15x^2 - 6x + 1$
12	$x^{12} - x^{11} - 11x^{10} + 10x^9 + 45x^8 - 36x^7 - 84x^6 + 56x^5 + 70x^4 - 35x^3 - 21x^2 + 6x + 1$
13	$x^{13} - x^{12} - 12x^{11} + 11x^{10} + 55x^9 - 45x^8 - 120x^7 + 84x^6 + 126x^5 - 70x^4 - 56x^3 + 21x^2 + 7x - 1$
14	$x^{14} - x^{13} - 13x^{12} + 12x^{11} + 66x^{10} - 55x^9 - 165x^8 + 120x^7 + 210x^6 - 126x^5 - 126x^4 +$ $+56x^3 + 28x^2 - 7x - 1$
15	$x^{15} - x^{14} - 14x^{13} + 13x^{12} + 78x^{11} - 66x^{10} - 220x^9 + 165x^8 + 330x^7 - 210x^6 - 252x^5 +$ $+126x^4 + 84x^3 - 28x^2 - 8x + 1$
16	$x^{16} - x^{15} - 15x^{14} + 14x^{13} + 91x^{12} - 78x^{11} - 286x^{10} + 220x^9 + 495x^8 - 330x^7 - 462x^6 + 252x^5 +$ $+210x^4 - 84x^3 - 36x^2 + 8x + 1$
17	$x^{17} - x^{16} - 16x^{15} + 15x^{14} + 105x^{13} - 91x^{12} - 364x^{11} + 286x^{10} + 715x^9 - 495x^8 - 792x^7 +$ $+462x^6 + 462x^5 - 210x^4 - 120x^3 + 36x^2 + 9x - 1$
18	$x^{18} - x^{17} - 17x^{16} + 16x^{15} + 120x^{14} - 105x^{13} - 455x^{12} + 364x^{11} + 1001x^{10} - 715x^9 - 1287x^8 +$ $+792x^7 + 924x^6 - 462x^5 - 330x^4 + 120x^3 + 45x^2 - 9x - 1$
19	$x^{19} - x^{18} - 18x^{17} + 17x^{16} + 136x^{15} - 120x^{14} - 560x^{13} + 455x^{12} + 1365x^{11} - 1001x^{10} - 2002x^9 +$ $+1287x^8 + 1716x^7 - 924x^6 - 792x^5 + 330x^4 + 165x^3 - 45x^2 - 10x + 1$
20	$x^{20} - x^{19} - 19x^{18} + 18x^{17} + 153x^{16} - 136x^{15} - 680x^{14} + 560x^{13} + 1820x^{12} - 1365x^{11} - 3003x^{10} +$ $+2002x^9 + 3003x^8 - 1716x^7 - 1716x^6 + 792x^5 + 495x^4 - 165x^3 - 55x^2 + 10x + 1$

Table 14: List of modified Chebyshev polynomials of the fourth kind for $1 \le n \le 20$

n	$W_n^*(x)$
1	$x+1$
2	x^2+x-1
3	x^3+x^2-2x-1
4	$x^4+x^3-3x^2-2x+1$
5	$x^5+x^4-4x^3-3x^2+3x+1$
6	$x^6+x^5-5x^4-4x^3+6x^2+3x-1$
7	$x^7+x^6-6x^5-5x^4+10x^3+6x^2-4x-1$
8	$x^8+x^7-7x^6-6x^5+15x^4+10x^3-10x^2-4x+1$
9	$x^9+x^8-8x^7-7x^6+21x^5+15x^4-20x^3-10x^2+5x+1$
10	$x^{10}+x^9-9x^8-8x^7+28x^6+21x^5-35x^4-20x^3+15x^2+5x-1$
11	$x^{11}+x^{10}-10x^9-9x^8+36x^7+28x^6-56x^5-35x^4+35x^3+15x^2-6x-1$
12	$x^{12}+x^{11}-11x^{10}-10x^9+45x^8+36x^7-84x^6-56x^5+70x^4+35x^3-21x^2-6x+1$
13	$x^{13}+x^{12}-12x^{11}-11x^{10}+55x^9+45x^8-120x^7-84x^6+126x^5+70x^4-56x^3-21x^2+7x+1$
14	$x^{14}+x^{13}-13x^{12}-12x^{11}+66x^{10}+55x^9-165x^8-120x^7+210x^6+126x^5-126x^4-$ $-56x^3+28x^2+7x-1$
15	$x^{15}+x^{14}-14x^{13}-13x^{12}+78x^{11}+66x^{10}-220x^9-165x^8+330x^7+210x^6-252x^5-$ $-126x^4+84x^3+28x^2-8x-1$
16	$x^{16}+x^{15}-15x^{14}-14x^{13}+91x^{12}+78x^{11}-286x^{10}-220x^9+495x^8+330x^7-462x^6-252x^5+$ $+210x^4+84x^3-36x^2-8x+1$
17	$x^{17}+x^{16}-16x^{15}-15x^{14}+105x^{13}+91x^{12}-364x^{11}-286x^{10}+715x^9+495x^8-792x^7-$ $-462x^6+462x^5+210x^4-120x^3-36x^2+9x+1$
18	$x^{18}+x^{17}-17x^{16}-16x^{15}+120x^{14}+105x^{13}-455x^{12}-364x^{11}+1001x^{10}+715x^9-1287x^8-$ $-792x^7+924x^6+462x^5-330x^4-120x^3+45x^2+9x-1$
19	$x^{19}+x^{18}-18x^{17}-17x^{16}+136x^{15}+120x^{14}-560x^{13}-455x^{12}+1365x^{11}+1001x^{10}-2002x^9-$ $-1287x^8+1716x^7+924x^6-792x^5-330x^4+165x^3+45x^2-10x-1$
20	$x^{20}+x^{19}-19x^{18}-18x^{17}+153x^{16}+136x^{15}-680x^{14}-560x^{13}+1820x^{12}+1365x^{11}-3003x^{10}-$ $-2002x^9+3003x^8+1716x^7-1716x^6-792x^5+495x^4+165x^3-55x^2-10x+1$

Table 15: Irreducible factorization of $T_n^*(x)$ for $1 \leq n \leq 20$

n	indirect	direct expression
1	$\Psi_4(x)$	x
2	$\Psi_8(x)$	x^2-2
3	$\Psi_4(x)\Psi_{12}(x)$	$x(x^2-3)$
4	$\Psi_{16}(x)$	x^4-4x^2+2
5	$\Psi_4(x)\Psi_{20}(x)$	$x(x^4-5x^2+5)$
6	$\Psi_8(x)\Psi_{24}(x)$	$(x^2-2)(x^4-4x^2+1)$
7	$\Psi_4(x)\Psi_{28}(x)$	$x(x^6-7x^4+14x^2-7)$
8	$\Psi_{32}(x)$	$x^8-8x^6+20x^4-16x^2+2$
9	$\Psi_4(x)\Psi_{12}(x)\Psi_{36}(x)$	$x(x^2-3)(x^6-6x^4+9x^2-3)$
10	$\Psi_8(x)\Psi_{40}(x)$	$(x^2-2)(x^8-8x^6+19x^4-12x^2+1)$
11	$\Psi_4(x)\Psi_{44}(x)$	$x(x^{10}-11x^8+44x^6-77x^4+55x^2-11)$
12	$\Psi_{16}(x)\Psi_{48}(x)$	$(x^4-4x^2+2)(x^8-8x^6+20x^4-16x^2+1)$
13	$\Psi_4(x)\Psi_{52}(x)$	$x(x^{12}-13x^{10}+65x^8-156x^6+182x^4-91x^2+13)$
14	$\Psi_8(x)\Psi_{56}(x)$	$(x^2-2)(x^{12}-12x^{10}+53x^8-104x^6+86x^4-24x^2+1)$
15	$\Psi_4(x)\Psi_{12}(x)\Psi_{20}(x)$ $\Psi_{60}(x)$	$x(x^2-3)(x^4-5x^2+5)(x^8-7x^6+14x^4-8x^2+1)$
16	$\Psi_{64}(x)$	$x^{16}-16x^{14}+104x^{12}-352x^{10}+660x^8-672x^6+336x^4-64x^2+2$
17	$\Psi_4(x)\Psi_{68}(x)$	$x(x^{16}-17x^{14}+119x^{12}-442x^{10}+935x^8-1122x^6+$ $+714x^4-204x^2+17)$
18	$\Psi_8(x)\Psi_{24}(x)\Psi_{72}(x)$	$(x^2-2)(x^4-4x^2+1)(x^{12}-12x^{10}+54x^8-112x^6+105x^4-36x^2+1)$
19	$\Psi_4(x)\Psi_{76}(x)$	$x(x^{18}-19x^{16}+152x^{14}-665x^{12}+1729x^{10}-2717x^8+$ $+2508x^6-1254x^4+285x^2-19)$
20	$\Psi_{16}(x)\Psi_{80}(x)$	$(x^4-4x^2+2)(x^{16}-16x^{14}+104x^{12}-352x^{10}+659x^8-664x^6+$ $+316x^4-48x^2+1)$

Table 16: Irreducible factorization of $T_n^*(x) - 2$ for $1 \leq n \leq 20$

n	indirect	direct expression
1	$[\Psi_1(x)]^2$	$x-2$
2	$[\Psi_1(x)]^2[\Psi_2(x)]^2$	$(x-2)(x+2)$
3	$[\Psi_1(x)]^2[\Psi_3(x)]^2$	$(x-2)(x+1)^2$
4	$[\Psi_1(x)]^2[\Psi_2(x)]^2[\Psi_4(x)]^2$	$(x-2)(x+2)x^2$
5	$[\Psi_1(x)]^2[\Psi_5(x)]^2$	$(x-2)(x^2+x-1)^2$
6	$[\Psi_1(x)]^2[\Psi_2(x)]^2[\Psi_3(x)]^2[\Psi_6(x)]^2$	$(x-2)(x+2)(x+1)^2(x-1)^2$
7	$[\Psi_1(x)]^2[\Psi_7(x)]^2$	$(x-2)(x^3+x^2-2x-1)^2$
8	$[\Psi_1(x)]^2[\Psi_2(x)]^2[\Psi_4(x)]^2[\Psi_8(x)]^2$	$(x-2)(x+2)x^2(x^2-2)^2$
9	$[\Psi_1(x)]^2[\Psi_3(x)]^2[\Psi_9(x)]^2$	$(x-2)(x+1)^2(x^3-3x+1)^2$
10	$[\Psi_1(x)]^2[\Psi_2(x)]^2[\Psi_5(x)]^2[\Psi_{10}(x)]^2$	$(x-2)(x+2)(x^2+x-1)^2(x^2-x-1)^2$
11	$[\Psi_1(x)]^2[\Psi_{11}(x)]^2$	$(x-2)(x^5+x^4-4x^3-3x^2+3x+1)^2$
12	$[\Psi_1(x)]^2[\Psi_2(x)]^2[\Psi_3(x)]^2[\Psi_4(x)]^2$ $[\Psi_6(x)]^2[\Psi_{12}(x)]^2$	$(x-2)(x+2)(x+1)^2x^2(x-1)^2(x^2-3)^2$
13	$[\Psi_1(x)]^2[\Psi_{13}(x)]^2$	$(x-2)(x^6+x^5-5x^4-4x^3+6x^2+3x-1)^2$
14	$[\Psi_1(x)]^2[\Psi_2(x)]^2[\Psi_7(x)]^2[\Psi_{14}(x)]^2$	$(x-2)(x+2)(x^3+x^2-2x-1)^2$ $(x^3-x^2-2x+1)^2$
15	$[\Psi_1(x)]^2[\Psi_3(x)]^2[\Psi_5(x)]^2[\Psi_{15}(x)]^2$	$(x-2)(x+1)^2(x^2+x-1)^2$ $(x^4-x^3-4x^2+4x+1)^2$
16	$[\Psi_1(x)]^2[\Psi_2(x)]^2[\Psi_4(x)]^2[\Psi_8(x)]^2[\Psi_{16}(x)]^2$	$(x-2)(x+2)x^2(x^2-2)^2(x^4-4x^2+2)^2$
17	$[\Psi_1(x)]^2[\Psi_{17}(x)]^2$	$(x-2)(x^8+x^7-7x^6-6x^5+15x^4+$ $+10x^3-10x^2-4x+1)^2$
18	$[\Psi_1(x)]^2[\Psi_2(x)]^2[\Psi_3(x)]^2[\Psi_6(x)]^2$ $[\Psi_9(x)]^2[\Psi_{18}(x)]^2$	$(x-2)(x+2)(x+1)^2(x-1)^2$ $(x^3-3x+1)^2(x^3-3x-1)^2$
19	$[\Psi_1(x)]^2[\Psi_{19}(x)]^2$	$(x-2)(x^9+x^8-8x^7-7x^6+21x^5+$ $+15x^4-20x^3-10x^2+5x+1)^2$
20	$[\Psi_1(x)]^2[\Psi_2(x)]^2[\Psi_4(x)]^2[\Psi_5(x)]^2$ $[\Psi_{10}(x)]^2[\Psi_{20}(x)]^2$	$(x-2)(x+2)x^2(x^2+x-1)^2(x^2-x-1)^2$ $(x^4-5x^2+5)^2$

Table 17: Irreducible factorization of $T_n^*(x)+2$ for $1 \le n \le 20$

n	indirect	direct expression
1	$[\Psi_2(x)]^2$	$x+2$
2	$[\Psi_4(x)]^2$	x^2
3	$[\Psi_2(x)]^2[\Psi_6(x)]^2$	$(x+2)(x-1)^2$
4	$[\Psi_8(x)]^2$	$(x^2-2)^2$
5	$[\Psi_2(x)]^2[\Psi_{10}(x)]^2$	$(x+2)(x^2-x-1)^2$
6	$[\Psi_4(x)]^2[\Psi_{12}(x)]^2$	$x^2(x^2-3)^2$
7	$[\Psi_2(x)]^2[\Psi_{14}(x)]^2$	$(x+2)(x^3-x^2-2x+1)^2$
8	$[\Psi_{16}(x)]^2$	$(x^4-4x^2+2)^2$
9	$[\Psi_2(x)]^2[\Psi_6(x)]^2[\Psi_{18}(x)]^2$	$(x+2)(x-1)^2(x^3-3x-1)^2$
10	$[\Psi_4(x)]^2[\Psi_{20}(x)]^2$	$x^2(x^4-5x^2+5)^2$
11	$[\Psi_2(x)]^2[\Psi_{22}(x)]^2$	$(x+2)(x^5-x^4-4x^3+3x^2+3x-1)^2$
12	$[\Psi_8(x)]^2[\Psi_{24}(x)]^2$	$(x^2-2)^2(x^4-4x^2+1)^2$
13	$[\Psi_2(x)]^2[\Psi_{26}(x)]^2$	$(x+2)(x^6-x^5-5x^4+4x^3+6x^2-3x-1)^2$
14	$[\Psi_4(x)]^2[\Psi_{28}(x)]^2$	$x^2(x^6-7x^4+14x^2-7)^2$
15	$[\Psi_2(x)]^2[\Psi_6(x)]^2[\Psi_{10}(x)]^2[\Psi_{30}(x)]^2$	$(x+2)(x-1)^2(x^2-x-1)^2(x^4+x^3-4x^2-4x+1)^2$
16	$[\Psi_{32}(x)]^2$	$(x^8-8x^6+20x^4-16x^2+2)^2$
17	$[\Psi_2(x)]^2[\Psi_{34}(x)]^2$	$(x+2)(x^8-x^7-7x^6+6x^5+15x^4-10x^3-10x^2+4x+1)^2$
18	$[\Psi_4(x)]^2[\Psi_{12}(x)]^2[\Psi_{36}(x)]^2$	$x^2(x^2-3)^2(x^6-6x^4+9x^2-3)^2$
19	$[\Psi_2(x)]^2[\Psi_{38}(x)]^2$	$(x+2)(x^9-x^8-8x^7+7x^6+$ $+21x^5-15x^4-20x^3+10x^2+5x-1)^2$
20	$[\Psi_8(x)]^2[\Psi_{40}(x)]^2$	$(x^2-2)^2(x^8-8x^6+19x^4-12x^2+1)^2$

Table 18: Irreducible factorization of $T_n^*(x) - 1$ for $1 \le n \le 20$

n	indirect	direct expression
1	$\Psi_6(x)$	$x - 1$
2	$\Psi_{12}(x)$	$x^2 - 3$
3	$\Psi_{18}(x)$	$x^3 - 3x - 1$
4	$\Psi_{24}(x)$	$x^4 - 4x^2 + 1$
5	$\Psi_6(x)\Psi_{30}(x)$	$(x-1)(x^4 + x^3 - 4x^2 - 4x + 1)$
6	$\Psi_{36}(x)$	$x^6 - 6x^4 + 9x^2 - 3$
7	$\Psi_6(x)\Psi_{42}(x)$	$(x-1)(x^6 + x^5 - 6x^4 - 6x^3 + 8x^2 + 8x + 1)$
8	$\Psi_{48}(x)$	$x^8 - 8x^6 + 20x^4 - 16x^2 + 1$
9	$\Psi_{54}(x)$	$x^9 - 9x^7 + 27x^5 - 30x^3 + 9x - 1$
10	$\Psi_{12}(x)\Psi_{60}(x)$	$(x^2-3)(x^8 - 7x^6 + 14x^4 - 8x^2 + 1)$
11	$\Psi_6(x)\Psi_{66}(x)$	$(x-1)(x^{10} + x^9 - 10x^8 - 10x^7 + 34x^6 + 34x^5 - 43x^4 -$ $-43x^3 + 12x^2 + 12x + 1)$
12	$\Psi_{72}(x)$	$x^{12} - 12x^{10} + 54x^8 - 112x^6 + 105x^4 - 36x^2 + 1$
13	$\Psi_6(x)\Psi_{78}(x)$	$(x-1)(x^{12} + x^{11} - 12x^{10} - 12x^9 + 53x^8 + 53x^7 -$ $-103x^6 - 103x^5 + 79x^4 + 79x^3 - 12x^2 - 12x + 1)$
14	$\Psi_{12}(x)\Psi_{84}(x)$	$(x^2-3)(x^{12} - 11x^{10} + 44x^8 - 78x^6 + 60x^4 - 16x^2 + 1)$
15	$\Psi_{18}(x)\Psi_{90}(x)$	$(x^3 - 3x - 1)(x^{12} - 12x^{10} + x^9 + 54x^8 - 9x^7 - 112x^6 + 27x^5$ $+105x^4 - 31x^3 - 36x^2 + 12x + 1)$
16	$\Psi_{96}(x)$	$x^{16} - 16x^{14} + 104x^{12} - 352x^{10} + 660x^8 - 672x^6 + 336x^4 - 64x^2 + 1$
17	$\Psi_6(x)\Psi_{102}(x)$	$(x-1)(x^{16} + x^{15} - 16x^{14} - 16x^{13} + 103x^{12} + 103x^{11} - 339x^{10} - 339x^9 +$ $+596x^8 + 596x^7 - 526x^6 - 526x^5 + 188x^4 + 188x^3 - 16x^2 - 16x + 1)$
18	$\Psi_{108}(x)$	$x^{18} - 18x^{16} + 135x^{14} - 546x^{12} + 1287x^{10} - 1782x^8 + 1386x^6 - 540x^4 + 81x^2 - 3$
19	$\Psi_6(x)\Psi_{114}(x)$	$(x-1)(x^{18} + x^{17} - 18x^{16} - 18x^{15} + 134x^{14} + 134x^{13} - 531x^{12} -$ $-531x^{11} + 1198x^{10} + 1198x^9 - 1519x^8 - 1519x^7 +$ $+989x^6 + 989x^5 - 265x^4 - 265x^3 + 20x^2 + 20x + 1)$
20	$\Psi_{24}(x)\Psi_{120}(x)$	$(x^4 - 4x^2 + 1)(x^{16} - 16x^{14} + 105x^{12} - 364x^{10} + 714x^8 - 784x^6 +$ $+440x^4 - 96x^2 + 1)$

Table 19: Irreducible factorization of $T_n^*(x)+1$ for $1 \leq n \leq 20$

n	indirect	direct expression
1	$\Psi_3(x)$	$x+1$
2	$\Psi_3(x)\Psi_6(x)$	$(x+1)(x-1)$
3	$\Psi_9(x)$	x^3-3x+1
4	$\Psi_3(x)\Psi_6(x)\Psi_{12}(x)$	$(x+1)(x-1)(x^2-3)$
5	$\Psi_3(x)\Psi_{15}(x)$	$(x+1)(x^4-x^3-4x^2+4x+1)$
6	$\Psi_9(x)\Psi_{18}(x)$	$(x^3-3x+1)(x^3-3x-1)$
7	$\Psi_3(x)\Psi_{21}(x)$	$(x+1)(x^6-x^5-6x^4+6x^3+8x^2-8x+1)$
8	$\Psi_3(x)\Psi_6(x)\Psi_{12}(x)\Psi_{24}(x)$	$(x+1)(x-1)(x^2-3)(x^4-4x^2+1)$
9	$\Psi_{27}(x)$	$x^9-9x^7+27x^5-30x^3+9x+1$
10	$\Psi_3(x)\Psi_6(x)\Psi_{15}(x)\Psi_{30}(x)$	$(x+1)(x-1)(x^4-x^3-4x^2+4x+1)(x^4+x^3-4x^2-4x+1)$
11	$\Psi_3(x)\Psi_{33}(x)$	$(x+1)(x^{10}-x^9-10x^8+10x^7+34x^6-34x^5-43x^4+$ $+43x^3+12x^2-12x+1)$
12	$\Psi_9(x)\Psi_{18}(x)\Psi_{36}(x)$	$(x^3-3x+1)(x^3-3x-1)(x^6-6x^4+9x^2-3)$
13	$\Psi_3(x)\Psi_{39}(x)$	$(x+1)(x^{12}-x^{11}-12x^{10}+12x^9+53x^8-53x^7-$ $-103x^6+103x^5+79x^4-79x^3-12x^2+12x+1)$
14	$\Psi_3(x)\Psi_6(x)\Psi_{21}(x)\Psi_{42}(x)$	$(x+1)(x-1)(x^6-x^5-6x^4+6x^3+8x^2-8x+1)$ $(x^6+x^5-6x^4-6x^3+8x^2+8x+1)$
15	$\Psi_9(x)\Psi_{45}(x)$	$(x^3-3x+1)(x^{12}-12x^{10}-x^9+54x^8+9x^7-112x^6-27x^5+$ $+105x^4+31x^3-36x^2-12x+1)$
16	$\Psi_3(x)\Psi_6(x)\Psi_{12}(x)\Psi_{24}(x)$ $\Psi_{48}(x)$	$(x+1)(x-1)(x^2-3)(x^4-4x^2+1)$ $(x^8-8x^6+20x^4-16x^2+1)$
17	$\Psi_3(x)\Psi_{51}(x)$	$(x+1)(x^{16}-x^{15}-16x^{14}+16x^{13}+103x^{12}-103x^{11}-$ $-339x^{10}+339x^9+596x^8-596x^7-526x^6+$ $+526x^5+188x^4-188x^3-16x^2+16x+1)$
18	$\Psi_{27}(x)\Psi_{54}(x)$	$(x^9-9x^7+27x^5-30x^3+9x+1)$ $(x^9-9x^7+27x^5-30x^3+9x-1)$
19	$\Psi_3(x)\Psi_{57}(x)$	$(x+1)(x^{18}-x^{17}-18x^{16}+18x^{15}+134x^{14}-134x^{13}-531x^{12}+$ $+531x^{11}+1198x^{10}-1198x^9-1519x^8+1519x^7+$ $+989x^6-989x^5-265x^4+265x^3+20x^2-20x+1)$
20	$\Psi_3(x)\Psi_6(x)\Psi_{12}(x)\Psi_{15}(x)$ $\Psi_{30}(x)\Psi_{60}(x)$	$(x+1)(x-1)(x^2-3)(x^4-x^3-4x^2+4x+1)$ $(x^4+x^3-4x^2-4x+1)(x^8-7x^6+14x^4-8x^2+1)$

Table 20: Irreducible factorization of $U_n^*(x)$ for $1 \le n \le 20$

n	indirect	direct expression
1	$\Psi_4(x)$	x
2	$\Psi_3(x)\Psi_6(x)$	$(x+1)(x-1)$
3	$\Psi_4(x)\Psi_8(x)$	$x(x^2-2)$
4	$\Psi_5(x)\Psi_{10}(x)$	$(x^2+x-1)(x^2-x-1)$
5	$\Psi_3(x)\Psi_4(x)\Psi_6(x)\Psi_{12}(x)$	$(x+1)x(x-1)(x^2-3)$
6	$\Psi_7(x)\Psi_{14}(x)$	$(x^3+x^2-2x-1)(x^3-x^2-2x+1)$
7	$\Psi_4(x)\Psi_8(x)\Psi_{16}(x)$	$x(x^2-2)(x^4-4x^2+2)$
8	$\Psi_3(x)\Psi_6(x)\Psi_9(x)\Psi_{18}(x)$	$(x+1)(x-1)(x^3-3x+1)(x^3-3x-1)$
9	$\Psi_4(x)\Psi_5(x)\Psi_{10}(x)\Psi_{20}(x)$	$x(x^2+x-1)(x^2-x-1)(x^4-5x^2+5)$
10	$\Psi_{11}(x)\Psi_{22}(x)$	$(x^5+x^4-4x^3-3x^2+3x+1)$ $(x^5-x^4-4x^3+3x^2+3x-1)$
11	$\Psi_3(x)\Psi_4(x)\Psi_6(x)\Psi_8(x)$ $\Psi_{12}(x)\Psi_{24}(x)$	$(x+1)x(x-1)(x^2-2)(x^2-3)(x^4-4x^2+1)$
12	$\Psi_{13}(x)\Psi_{26}(x)$	$(x^6+x^5-5x^4-4x^3+6x^2+3x-1)$ $(x^6-x^5-5x^4+4x^3+6x^2-3x-1)$
13	$\Psi_4(x)\Psi_7(x)\Psi_{14}(x)\Psi_{28}(x)$	$x(x^3+x^2-2x-1)(x^3-x^2-2x+1)$ $(x^6-7x^4+14x^2-7)$
14	$\Psi_3(x)\Psi_5(x)\Psi_6(x)\Psi_{10}(x)$ $\Psi_{15}(x)\Psi_{30}(x)$	$(x+1)(x^2+x-1)(x-1)(x^2-x-1)$ $(x^4-x^3-4x^2+4x+1)(x^4+x^3-4x^2-4x+1)$
15	$\Psi_4(x)\Psi_8(x)\Psi_{16}(x)\Psi_{32}(x)$	$x(x^2-2)(x^4-4x^2+2)(x^8-8x^6+20x^4-16x^2+2)$
16	$\Psi_{17}(x)\Psi_{34}(x)$	$(x^8+x^7-7x^6-6x^5+15x^4+10x^3-10x^2-4x+1)$ $(x^8-x^7-7x^6+6x^5+15x^4-10x^3-10x^2+4x+1)$
17	$\Psi_3(x)\Psi_4(x)\Psi_6(x)\Psi_9(x)\Psi_{12}(x)$ $\Psi_{18}(x)\Psi_{36}(x)$	$(x+1)x(x-1)(x^3-3x+1)(x^2-3)$ $(x^3-3x-1)(x^6-6x^4+9x^2-3)$
18	$\Psi_{19}(x)\Psi_{38}(x)$	$(x^9+x^8-8x^7-7x^6+21x^5+15x^4-20x^3-10x^2+5x+1)$ $(x^9-x^8-8x^7+7x^6+21x^5-15x^4-20x^3+10x^2+5x-1)$
19	$\Psi_4(x)\Psi_5(x)\Psi_8(x)\Psi_{10}(x)$ $\Psi_{20}(x)\Psi_{40}(x)$	$x(x^2+x-1)(x^2-2)(x^2-x-1)(x^4-5x^2+5)$ $(x^8-8x^6+19x^4-12x^2+1)$
20	$\Psi_3(x)\Psi_6(x)\Psi_7(x)\Psi_{14}(x)$ $\Psi_{21}(x)\Psi_{42}(x)$	$(x+1)(x-1)(x^3+x^2-2x-1)(x^3-x^2-2x+1)$ $(x^6-x^5-6x^4+6x^3+8x^2-8x+1)$ $(x^6+x^5-6x^4-6x^3+8x^2+8x+1)$

Table 21: Irreducible factorization of $U_n^*(x) - 1$ for $1 \leq n \leq 20$

n	indirect	direct expression
1	$\Psi_6(x)$	$x-1$
2	$\Psi_8(x)$	x^2-2
3	$\Psi_3(x)\Psi_{10}(x)$	$(x+1)(x^2-x-1)$
4	$[\Psi_4(x)]^2\Psi_{12}(x)$	$x^2(x^2-3)$
5	$\Psi_5(x)\Psi_{14}(x)$	$(x^2+x-1)(x^3-x^2-2x+1)$
6	$\Psi_3(x)\Psi_6(x)\Psi_{16}(x)$	$(x+1)(x-1)(x^4-4x^2+2)$
7	$\Psi_6(x)\Psi_7(x)\Psi_{18}(x)$	$(x-1)(x^3+x^2-2x-1)(x^3-3x-1)$
8	$[\Psi_4(x)]^2\Psi_8(x)\Psi_{20}(x)$	$x^2(x^2-2)(x^4-5x^2+5)$
9	$\Psi_3(x)\Psi_9(x)\Psi_{22}(x)$	$(x+1)(x^3-3x+1)(x^5-x^4-4x^3+3x^2+3x-1)$
10	$\Psi_5(x)\Psi_8(x)\Psi_{10}(x)\Psi_{24}(x)$	$(x^2+x-1)(x^2-2)(x^2-x-1)(x^4-4x^2+1)$
11	$\Psi_{11}(x)\Psi_{26}(x)$	$(x^5+x^4-4x^3-3x^2+3x+1)$ $(x^6-x^5-5x^4+4x^3+6x^2-3x-1)$
12	$\Psi_3(x)[\Psi_4(x)]^2\Psi_6(x)\Psi_{12}(x)\Psi_{28}(x)$	$(x+1)x^2(x-1)(x^2-3)(x^6-7x^4+14x^2-7)$
13	$\Psi_6(x)\Psi_{10}(x)\Psi_{13}(x)\Psi_{30}(x)$	$(x-1)(x^2-x-1)(x^6+x^5-5x^4-4x^3+6x^2+3x-1)$ $(x^4+x^3-4x^2-4x+1)$
14	$\Psi_7(x)\Psi_{14}(x)\Psi_{32}(x)$	$(x^3+x^2-2x-1)(x^3-x^2-2x+1)$ $(x^8-8x^6+20x^4-16x^2+2)$
15	$\Psi_3(x)\Psi_5(x)\Psi_{15}(x)\Psi_{34}(x)$	$(x+1)(x^2+x-1)(x^4-x^3-4x^2+4x+1)(x^8-x^7-$ $-7x^6+6x^5+15x^4-10x^3-10x^2+4x+1)$
16	$[\Psi_4(x)]^2\Psi_8(x)\Psi_{12}(x)\Psi_{16}(x)\Psi_{36}(x)$	$x^2(x^2-2)(x^2-3)(x^4-4x^2+2)(x^6-6x^4+9x^2-3)$
17	$\Psi_{17}(x)\Psi_{38}(x)$	$(x^8+x^7-7x^6-6x^5+15x^4+10x^3-10x^2-4x+1)$ $(x^9-x^8-8x^7+7x^6+21x^5-15x^4-20x^3+10x^2+5x-1)$
18	$\Psi_3(x)\Psi_6(x)\Psi_8(x)\Psi_9(x)$ $\Psi_{18}(x)\Psi_{40}(x)$	$(x+1)(x-1)(x^2-2)(x^3-3x+1)(x^3-3x-1)$ $(x^8-8x^6+19x^4-12x^2+1)$
19	$\Psi_6(x)\Psi_{14}(x)\Psi_{19}(x)\Psi_{42}(x)$	$(x-1)(x^3-x^2-2x+1)$ $(x^9+x^8-8x^7-7x^6+21x^5+15x^4-20x^3-10x^2+5x+1)$ $(x^6+x^5-6x^4-6x^3+8x^2+8x+1)$
20	$[\Psi_4(x)]^2\Psi_5(x)\Psi_{10}(x)\Psi_{20}(x)\Psi_{44}(x)$	$x^2(x^2+x-1)(x^2-x-1)(x^4-5x^2+5)$ $(x^{10}-11x^8+44x^6-77x^4+55x^2-11)$

Table 22: Irreducible factorization of $U_n^*(x)+1$ for $1 \le n \le 20$

n	indirect	direct expression
1	$\Psi_3(x)$	$x+1$
2	$[\Psi_4(x)]^2$	x^2
3	$\Psi_5(x)\Psi_6(x)$	$(x^2+x-1)(x-1)$
4	$\Psi_3(x)\Psi_6(x)\Psi_8(x)$	$(x+1)(x-1)(x^2-2)$
5	$\Psi_7(x)\Psi_{10}(x)$	$(x^3+x^2-2x-1)(x^2-x-1)$
6	$[\Psi_4(x)]^2\Psi_8(x)\Psi_{12}(x)$	$x^2(x^2-2)(x^2-3)$
7	$\Psi_3(x)\Psi_9(x)\Psi_{14}(x)$	$(x+1)(x^3-3x+1)(x^3-x^2-2x+1)$
8	$\Psi_5(x)\Psi_{10}(x)\Psi_{16}(x)$	$(x^2+x-1)(x^2-x-1)(x^4-4x^2+2)$
9	$\Psi_6(x)\Psi_{11}(x)\Psi_{18}(x)$	$(x-1)(x^5+x^4-4x^3-3x^2+3x+1)(x^3-3x-1)$
10	$\Psi_3(x)[\Psi_4(x)]^2\Psi_6(x)$ $\Psi_{12}(x)\Psi_{20}(x)$	$(x+1)x^2(x-1)(x^2-3)(x^4-5x^2+5)$
11	$\Psi_{13}(x)\Psi_{22}(x)$	$(x^6+x^5-5x^4-4x^3+6x^2+3x-1)$ $(x^5-x^4-4x^3+3x^2+3x-1)$
12	$\Psi_7(x)\Psi_8(x)\Psi_{14}(x)\Psi_{24}(x)$	$(x^3+x^2-2x-1)(x^2-2)(x^3-x^2-2x+1)(x^4-4x^2+1)$
13	$\Psi_3(x)\Psi_5(x)\Psi_{15}(x)\Psi_{26}(x)$	$(x+1)(x^2+x-1)(x^4-x^3-4x^2+4x+1)$ $(x^6-x^5-5x^4+4x^3+6x^2-3x-1)$
14	$[\Psi_4(x)]^2\Psi_8(x)\Psi_{16}(x)\Psi_{28}(x)$	$x^2(x^2-2)(x^4-4x^2+2)(x^6-7x^4+14x^2-7)$
15	$\Psi_6(x)\Psi_{10}(x)\Psi_{17}(x)\Psi_{30}(x)$	$(x-1)(x^2-x-1)(x^8+x^7-7x^6-6x^5+15x^4+10x^3-)$ $-10x^2-4x+1)(x^4+x^3-4x^2-4x+1)$
16	$\Psi_3(x)\Psi_6(x)\Psi_9(x)\Psi_{18}(x)$ $\Psi_{32}(x)$	$(x+1)(x-1)(x^3-3x+1)(x^3-3x-1)$ $(x^8-8x^6+20x^4-16x^2+2)$
17	$\Psi_{19}(x)\Psi_{34}(x)$	$(x^9+x^8-8x^7-7x^6+21x^5+15x^4-20x^3-10x^2+5x+1)$ $(x^8-x^7-7x^6+6x^5+15x^4-10x^3-10x^2+4x+1)$
18	$[\Psi_4(x)]^2)\Psi_5(x)\Psi_{10}(x)\Psi_{12}(x)$ $\Psi_{20}(x)\Psi_{36}(x)$	$x^2(x^2+x-1)(x^2-x-1)(x^2-3)$ $(x^4-5x^2+5)(x^6-6x^4+9x^2-3)$
19	$\Psi_3(x)\Psi_7(x)\Psi_{21}(x)\Psi_{38}(x)$	$(x+1)(x^3+x^2-2x-1)(x^6-x^5-6x^4+6x^3+8x^2-8x+1)$ $(x^9-x^8-8x^7+7x^6+21x^5-15x^4-20x^3+10x^2+5x-1)$
20	$\Psi_8(x)\Psi_{11}(x)\Psi_{22}(x)\Psi_{40}(x)$	$(x^2-2)(x^5+x^4-4x^3-3x^2+3x+1)(x^5-x^4-4x^3+3x^2+$ $+3x-1)(x^8-8x^6+19x^4-12x^2+1)$

Table 23: Irreducible factorization of $V_n^*(x)$ for $1 \leq n \leq 20$

n	indirect	direct expression
1	$\Psi_6(x)$	$x-1$
2	$\Psi_{10}(x)$	x^2-x-1
3	$\Psi_{14}(x)$	x^3-x^2-2x+1
4	$\Psi_6(x)\Psi_{18}(x)$	$(x-1)(x^3-3x-1)$
5	$\Psi_{22}(x)$	$x^5-x^4-4x^3+3x^2+3x-1$
6	$\Psi_{26}(x)$	$x^6-x^5-5x^4+4x^3+6x^2-3x-1$
7	$\Psi_6(x)\Psi_{10}(x)\Psi_{30}(x)$	$(x-1)(x^2-x-1)(x^4+x^3-4x^2-4x+1)$
8	$\Psi_{34}(x)$	$x^8-x^7-7x^6+6x^5+15x^4-10x^3-10x^2+4x+1$
9	$\Psi_{38}(x)$	$x^9-x^8-8x^7+7x^6+21x^5-15x^4-20x^3+10x^2+5x-1$
10	$\Psi_6(x)\Psi_{14}(x)\Psi_{42}(x)$	$(x-1)(x^3-x^2-2x+1)(x^6+x^5-6x^4-6x^3+8x^2+8x+1)$
11	$\Psi_{46}(x)$	$x^{11}-x^{10}-10x^9+9x^8+36x^7-28x^6-56x^5+35x^4+35x^3-$ $-15x^2-6x+1$
12	$\Psi_{10}(x)\Psi_{50}(x)$	$(x^2-x-1)(x^{10}-10x^8+35x^6-x^5-50x^4+5x^3+25x^2-$ $-5x-1)$
13	$\Psi_6(x)\Psi_{18}(x)\Psi_{54}(x)$	$(x-1)(x^3-3x-1)(x^9-9x^7+27x^5-30x^3+9x-1)$
14	$\Psi_{58}(x)$	$x^{14}-x^{13}-13x^{12}+12x^{11}+66x^{10}-55x^9-165x^8+120x^7+$ $+210x^6-126x^5-126x^4+56x^3+28x^2-7x-1$
15	$\Psi_{62}(x)$	$x^{15}-x^{14}-14x^{13}+13x^{12}+78x^{11}-66x^{10}-220x^9+165x^8+$ $+330x^7-210x^6-252x^5+126x^4+84x^3-28x^2-8x+1$
16	$\Psi_6(x)\Psi_{22}(x)\Psi_{66}(x)$	$(x-1)(x^5-x^4-4x^3+3x^2+3x-1)(x^{10}+x^9-10x^8-$ $-10x^7+34x^6+34x^5-43x^4-43x^3+12x^2+12x+1)$
17	$\Psi_{10}(x)\Psi_{14}(x)\Psi_{70}(x)$	$(x^2-x-1)(x^3-x^2-2x+1)(x^{12}+x^{11}-12x^{10}-11x^9+$ $+54x^8+43x^7-113x^6-71x^5+110x^4+$ $+46x^3-40x^2-8x+1)$
18	$\Psi_{74}(x)$	$x^{18}-x^{17}-17x^{16}+16x^{15}+120x^{14}-105x^{13}-455x^{12}+$ $+364x^{11}+1001x^{10}-715x^9-1287x^8+792x^7+$ $+924x^6-462x^5-330x^4+120x^3+45x^2-9x-1$
19	$\Psi_6(x)\Psi_{26}(x)\Psi_{78}(x)$	$(x-1)(x^6-x^5-5x^4+4x^3+6x^2-3x-1)(x^{12}+x^{11}-$ $-12x^{10}-12x^9+53x^8+53x^7-103x^6-103x^5+$ $+79x^4+79x^3-12x^2-12x+1)$
20	$\Psi_{82}(x)$	$x^{20}-x^{19}-19x^{18}+18x^{17}+153x^{16}-136x^{15}-680x^{14}+$ $+560x^{13}+1820x^{12}-1365x^{11}-3003x^{10}+2002x^9+$ $+3003x^8-1716x^7-1716x^6+792x^5+495x^4-165x^3-$ $-55x^2+10x+1$

Table 24: Irreducible factorization of $V_n^*(x) - 1$ for $1 \le n \le 20$

n	indirect	direct expression
1	$[\Psi_1(x)]^2$	$x-2$
2	$[\Psi_1(x)]^2\Psi_3(x)$	$(x-2)(x+1)$
3	$[\Psi_1(x)]^2\Psi_3(x)\Psi_4(x)$	$(x-2)(x+1)x$
4	$[\Psi_1(x)]^2\Psi_4(x)\Psi_5(x)$	$(x-2)x(x^2+x-1)$
5	$[\Psi_1(x)]^2\Psi_3(x)\Psi_5(x)\Psi_6(x)$	$(x-2)(x+1)(x^2+x-1)(x-1)$
6	$[\Psi_1(x)]^2\Psi_3(x)\Psi_6(x)\Psi_7(x)$	$(x-2)(x+1)(x-1)(x^3+x^2-2x-1)$
7	$[\Psi_1(x)]^2\Psi_4(x)\Psi_7(x)\Psi_8(x)$	$(x-2)x(x^3+x^2-2x-1)(x^2-2)$
8	$[\Psi_1(x)]^2\Psi_3(x)\Psi_4(x)\Psi_8(x)\Psi_9(x)$	$(x-2)(x+1)x(x^2-2)(x^3-3x+1)$
9	$[\Psi_1(x)]^2\Psi_3(x)\Psi_5(x)\Psi_9(x)\Psi_{10}(x)$	$(x-2)(x+1)(x^2+x-1)(x^3-3x+1)(x^2-x-1)$
10	$[\Psi_1(x)]^2\Psi_5(x)\Psi_{10}(x)\Psi_{11}(x)$	$(x-2)(x^2+x-1)(x^2-x-1)(x^5+x^4-4x^3-3x^2+3x+1)$
11	$[\Psi_1(x)]^2\Psi_3(x)\Psi_4(x)\Psi_6(x)$ $\Psi_{11}(x)\Psi_{12}(x)$	$(x-2)(x+1)x(x-1)(x^5+x^4-4x^3-3x^2+3x+1)(x^2-3)$
12	$[\Psi_1(x)]^2\Psi_3(x)\Psi_4(x)\Psi_6(x)$ $\Psi_{12}(x)\Psi_{13}(x)$	$(x-2)(x+1)x(x-1)(x^2-3)$ $(x^6+x^5-5x^4-4x^3+6x^2+3x-1)$
13	$[\Psi_1(x)]^2\Psi_7(x)\Psi_{13}(x)\Psi_{14}(x)$	$(x-2)(x^3+x^2-2x-1)(x^6+x^5-5x^4-4x^3+$ $+6x^2+3x-1)(x^3-x^2-2x+1)$
14	$[\Psi_1(x)]^2\Psi_3(x)\Psi_5(x)\Psi_7(x)\Psi_{14}(x)$ $\Psi_{15}(x)$	$(x-2)(x+1)(x^2+x-1)(x^3+x^2-2x-1)$ $(x^3-x^2-2x+1)(x^4-x^3-4x^2+4x+1)$
15	$[\Psi_1(x)]^2\Psi_3(x)\Psi_4(x)\Psi_5(x)\Psi_8(x)$ $\Psi_{15}(x)\Psi_{16}(x)$	$(x-2)(x+1)x(x^2+x-1)(x^2-2)(x^4-x^3-4x^2+4x+1)$ (x^4-4x^2+2)
16	$[\Psi_1(x)]^2\Psi_4(x)\Psi_8(x)\Psi_{16}(x)$ $\Psi_{17}(x)$	$(x-2)x(x^2-2)(x^4-4x^2+2)$ $(x^8+x^7-7x^6-6x^5+15x^4+10x^3-10x^2-4x+1)$
17	$[\Psi_1(x)]^2\Psi_3(x)\Psi_6(x)\Psi_9(x)\Psi_{17}(x)$ $\Psi_{18}(x)$	$(x-2)(x+1)(x-1)(x^3-3x+1)(x^8+x^7-7x^6-6x^5+$ $+15x^4+10x^3-10x^2-4x+1)(x^3-3x-1)$
18	$[\Psi_1(x)]^2\Psi_3(x)\Psi_6(x)\Psi_9(x)\Psi_{18}(x)$ $\Psi_{19}(x)$	$(x-2)(x+1)(x-1)(x^3-3x+1)(x^3-3x-1)(x^9+x^8-$ $-8x^7-7x^6+21x^5+15x^4-20x^3-10x^2+5x+1)$
19	$[\Psi_1(x)]^2\Psi_4(x)\Psi_5(x)\Psi_{10}(x)$ $\Psi_{19}(x)\Psi_{20}(x)$	$(x-2)x(x^2+x-1)(x^2-x-1)(x^9+x^8-8x^7-7x^6+$ $+21x^5+15x^4-20x^3-10x^2+5x+1)(x^4-5x^2+5)$
20	$[\Psi_1(x)]^2\Psi_3(x)\Psi_4(x)\Psi_5(x)\Psi_7(x)$ $\Psi_{10}(x)\Psi_{20}(x)\Psi_{21}(x)$	$(x-2)(x+1)x(x^2+x-1)(x^3+x^2-2x-1)(x^2-x-1)$ $(x^4-5x^2+5)(x^6-x^5-6x^4+6x^3+8x^2-8x+1)$

Table 25: Irreducible factorization of $V_n^*(x)+1$ for $1 \le n \le 20$

n	indirect	direct expression
1	$\Psi_4(x)$	x
2	$\Psi_4(x)\Psi_6(x)$	$x(x-1)$
3	$\Psi_6(x)\Psi_8(x)$	$(x-1)(x^2-2)$
4	$\Psi_8(x)\Psi_{10}(x)$	$(x^2-2)(x^2-x-1)$
5	$\Psi_4(x)\Psi_{10}(x)\Psi_{12}(x)$	$x(x^2-x-1)(x^2-3)$
6	$\Psi_4(x)\Psi_{12}(x)\Psi_{14}(x)$	$x(x^2-3)(x^3-x^2-2x+1)$
7	$\Psi_{14}(x)\Psi_{16}(x)$	$(x^3-x^2-2x+1)(x^4-4x^2+2)$
8	$\Psi_6(x)\Psi_{16}(x)\Psi_{18}(x)$	$(x-1)(x^4-4x^2+2)(x^3-3x-1)$
9	$\Psi_4(x)\Psi_6(x)\Psi_{18}(x)\Psi_{20}(x)$	$x(x-1)(x^3-3x-1)(x^4-5x^2+5)$
10	$\Psi_4(x)\Psi_{20}(x)\Psi_{22}(x)$	$x(x^4-5x^2+5)(x^5-x^4-4x^3+3x^2+3x-1)$
11	$\Psi_8(x)\Psi_{22}(x)\Psi_{24}(x)$	$(x^2-2)(x^5-x^4-4x^3+3x^2+3x-1)(x^4-4x^2+1)$
12	$\Psi_8(x)\Psi_{24}(x)\Psi_{26}(x)$	$(x^2-2)(x^4-4x^2+1)(x^6-x^5-5x^4+4x^3+6x^2-3x-1)$
13	$\Psi_4(x)\Psi_{26}(x)\Psi_{28}(x)$	$x(x^6-x^5-5x^4+4x^3+6x^2-3x-1)(x^6-7x^4+14x^2-7)$
14	$\Psi_4(x)\Psi_6(x)\Psi_{10}(x)\Psi_{28}(x)$ $\Psi_{30}(x)$	$x(x-1)(x^2-x-1)(x^6-7x^4+14x^2-7)$ $(x^4+x^3-4x^2-4x+1)$
15	$\Psi_6(x)\Psi_{10}(x)\Psi_{30}(x)$ $\Psi_{32}(x)$	$(x-1)(x^2-x-1)(x^4+x^3-4x^2-4x+1)$ $(x^8-8x^6+20x^4-16x^2+2)$
16	$\Psi_{32}(x)\Psi_{34}(x)$	$(x^8-8x^6+20x^4-16x^2+2)$ $(x^8-x^7-7x^6+6x^5+15x^4-10x^3-10x^2+4x+1)$
17	$\Psi_4(x)\Psi_{12}(x)\Psi_{34}(x)$ $\Psi_{36}(x)$	$x(x^2-3)(x^8-x^7-7x^6+6x^5+15x^4-10x^3-10x^2+4x+1)$ $(x^6-6x^4+9x^2-3)$
18	$\Psi_4(x)\Psi_{12}(x)\Psi_{36}(x)$ $\Psi_{38}(x)$	$x(x^2-3)(x^6-6x^4+9x^2-3)$ $(x^9-x^8-8x^7+7x^6+21x^5-15x^4-20x^3+10x^2+5x-1)$
19	$\Psi_8(x)\Psi_{38}(x)\Psi_{40}(x)$	$(x^2-2)(x^9-x^8-8x^7+7x^6+21x^5-15x^4-20x^3+$ $+10x^2+5x-1)(x^8-8x^6+19x^4-12x^2+1)$
20	$\Psi_6(x)\Psi_8(x)\Psi_{14}(x)\Psi_{40}(x)$ $\Psi_{42}(x)$	$(x-1)(x^2-2)(x^3-x^2-2x+1)(x^8-8x^6+19x^4-12x^2+1)$ $(x^6+x^5-6x^4-6x^3+8x^2+8x+1)$

Table 26: Irreducible factorization of $W_n^*(x)$ for $1 \le n \le 20$

n	indirect	direct expression
1	$\Psi_3(x)$	$x+1$
2	$\Psi_5(x)$	x^2+x-1
3	$\Psi_7(x)$	x^3+x^2-2x-1
4	$\Psi_3(x)\Psi_9(x)$	$(x+1)(x^3-3x+1)$
5	$\Psi_{11}(x)$	$x^5+x^4-4x^3-3x^2+3x+1$
6	$\Psi_{13}(x)$	$x^6+x^5-5x^4-4x^3+6x^2+3x-1$
7	$\Psi_3(x)\Psi_5(x)\Psi_{15}(x)$	$(x+1)(x^2+x-1)(x^4-x^3-4x^2+4x+1)$
8	$\Psi_{17}(x)$	$x^8+x^7-7x^6-6x^5+15x^4+10x^3-10x^2-4x+1$
9	$\Psi_{19}(x)$	$x^9+x^8-8x^7-7x^6+21x^5+15x^4-20x^3-10x^2+5x+1$
10	$\Psi_3(x)\Psi_7(x)\Psi_{21}(x)$	$(x+1)(x^3+x^2-2x-1)(x^6-x^5-6x^4+6x^3+8x^2-8x+1)$
11	$\Psi_{23}(x)$	$x^{11}+x^{10}-10x^9-9x^8+36x^7+28x^6-56x^5-35x^4+35x^3+$ $+15x^2-6x-1$
12	$\Psi_5(x)\Psi_{25}(x)$	$(x^2+x-1)(x^{10}-10x^8+35x^6+x^5-50x^4-5x^3+25x^2+$ $+5x-1)$
13	$\Psi_3(x)\Psi_9(x)\Psi_{27}(x)$	$(x+1)(x^3-3x+1)(x^9-9x^7+27x^5-30x^3+9x+1)$
14	$\Psi_{29}(x)$	$x^{14}+x^{13}-13x^{12}-12x^{11}+66x^{10}+55x^9-165x^8-120x^7+$ $+210x^6+126x^5-126x^4-56x^3+28x^2+7x-1$
15	$\Psi_{31}(x)$	$x^{15}+x^{14}-14x^{13}-13x^{12}+78x^{11}+66x^{10}-220x^9-165x^8+$ $+330x^7+210x^6-252x^5-126x^4+84x^3+28x^2-8x-1$
16	$\Psi_3(x)\Psi_{11}(x)\Psi_{33}(x)$	$(x+1)(x^5+x^4-4x^3-3x^2+3x+1)(x^{10}-x^9-10x^8+$ $+10x^7+34x^6-34x^5-43x^4+43x^3+12x^2-12x+1)$
17	$\Psi_5(x)\Psi_7(x)\Psi_{35}(x)$	$(x^2+x-1)(x^3+x^2-2x-1)(x^{12}-x^{11}-12x^{10}+11x^9+$ $+54x^8-43x^7-113x^6+71x^5+110x^4-$ $-46x^3-40x^2+8x+1)$
18	$\Psi_{37}(x)$	$x^{18}+x^{17}-17x^{16}-16x^{15}+120x^{14}+105x^{13}-455x^{12}-$ $-364x^{11}+1001x^{10}+715x^9-1287x^8-792x^7+$ $+924x^6+462x^5-330x^4-120x^3+45x^2+9x-1$
19	$\Psi_3(x)\Psi_{13}(x)\Psi_{39}(x)$	$(x+1)(x^6+x^5-5x^4-4x^3+6x^2+3x-1)(x^{12}-x^{11}-$ $-12x^{10}+12x^9+53x^8-53x^7-103x^6+103x^5+$ $+79x^4-79x^3-12x^2+12x+1)$
20	$\Psi_{41}(x)$	$x^{20}+x^{19}-19x^{18}-18x^{17}+153x^{16}+136x^{15}-680x^{14}-$ $-560x^{13}+1820x^{12}+1365x^{11}-3003x^{10}-2002x^9+$ $+3003x^8+1716x^7-1716x^6-792x^5+495x^4+165x^3-$ $-55x^2-10x+1$

Table 27: Irreducible factorization of $W_n^*(x) - 1$ for $1 \le n \le 20$

n	indirect	direct expression
1	$\Psi_4(x)$	x
2	$[\Psi_2(x)]^2\Psi_6(x)$	$(x+2)(x-1)$
3	$\Psi_3(x)\Psi_8(x)$	$(x+1)(x^2-2)$
4	$[\Psi_2(x)]^2\Psi_4(x)\Psi_{10}(x)$	$(x+2)x(x^2-x-1)$
5	$\Psi_4(x)\Psi_5(x)\Psi_{12}(x)$	$x(x^2+x-1)(x^2-3)$
6	$[\Psi_2(x)]^2\Psi_3(x)\Psi_6(x)\Psi_{14}(x)$	$(x+2)(x+1)(x-1)(x^3-x^2-2x+1)$
7	$\Psi_7(x)\Psi_{16}(x)$	$(x^3+x^2-2x-1)(x^4-4x^2+2)$
8	$[\Psi_2(x)]^2\Psi_4(x)\Psi_6(x)\Psi_8(x)\Psi_{18}(x)$	$(x+2)x(x-1)(x^2-2)(x^3-3x-1)$
9	$\Psi_3(x)\Psi_4(x)\Psi_9(x)\Psi_{20}(x)$	$(x+1)x(x^3-3x+1)(x^4-5x^2+5)$
10	$[\Psi_2(x)]^2\Psi_5(x)\Psi_{10}(x)\Psi_{22}(x)$	$(x+2)(x^2+x-1)(x^2-x-1)(x^5-x^4-4x^3+3x^2+3x-1)$
11	$\Psi_8(x)\Psi_{11}(x)\Psi_{24}(x)$	$(x^2-2)(x^5+x^4-4x^3-3x^2+3x+1)(x^4-4x^2+1)$
12	$[\Psi_2(x)]^2\Psi_3(x)\Psi_4(x)\Psi_6(x)\Psi_{12}(x)$ $\Psi_{26}(x)$	$(x+2)(x+1)x(x-1)(x^2-3)$ $(x^6-x^5-5x^4+4x^3+6x^2-3x-1)$
13	$\Psi_4(x)\Psi_{13}(x)\Psi_{28}(x)$	$x(x^6+x^5-5x^4-4x^3+6x^2+3x-1)(x^6-7x^4+14x^2-7)$
14	$[\Psi_2(x)]^2\Psi_6(x)\Psi_7(x)\Psi_{10}(x)\Psi_{14}(x)$ $\Psi_{30}(x)$	$(x+2)(x-1)(x^3+x^2-2x-1)(x^2-x-1)(x^3-x^2-2x+1)$ $(x^4+x^3-4x^2-4x+1)$
15	$\Psi_3(x)\Psi_5(x)\Psi_{15}(x)\Psi_{32}(x)$	$(x+1)(x^2+x-1)(x^4-x^3-4x^2+4x+1)$ $(x^8-8x^6+20x^4-16x^2+2)$
16	$[\Psi_2(x)]^2\Psi_4(x)\Psi_8(x)\Psi_{16}(x)\Psi_{34}(x)$	$(x+2)x(x^2-2)(x^4-4x^2+2)$ $(x^8-x^7-7x^6+6x^5+15x^4-10x^3-10x^2+4x+1)$
17	$\Psi_4(x)\Psi_{12}(x)\Psi_{17}(x)\Psi_{36}(x)$	$x(x^2-3)(x^8+x^7-7x^6-6x^5+15x^4+10x^3-10x^2-4x+1)$ $(x^6-6x^4+9x^2-3)$
18	$[\Psi_2(x)]^2\Psi_3(x)\Psi_6(x)\Psi_9(x)\Psi_{18}(x)$ $\Psi_{38}(x)$	$(x+2)(x+1)(x-1)(x^3-3x+1)(x^3-3x-1)(x^9-x^8-$ $-8x^7+7x^6+21x^5-15x^4-20x^3+10x^2+5x-1)$
19	$\Psi_8(x)\Psi_{19}(x)\Psi_{40}(x)$	$(x^2-2)(x^9+x^8-8x^7-7x^6+21x^5+15x^4-20x^3-$ $-10x^2+5x+1)(x^8-8x^6+19x^4-12x^2+1)$
20	$[\Psi_2(x)]^2\Psi_4(x)\Psi_5(x)\Psi_6(x)\Psi_{10}(x)$ $\Psi_{14}(x)\Psi_{20}(x)\Psi_{42}(x)$	$(x+2)x(x^2+x-1)(x-1)(x^2-x-1)(x^3-x^2-2x+1)$ $(x^4-5x^2+5)(x^6+x^5-6x^4-6x^3+8x^2+8x+1)$

Table 28: Irreducible factorization of $W_n^*(x)+1$ for $1 \leq n \leq 20$

n	indirect	direct expression
1	$[\Psi_2(x)]^2$	$x+2$
2	$\Psi_3(x)\Psi_4(x)$	$(x+1)x$
3	$[\Psi_2(x)]^2\Psi_4(x)\Psi_6(x)$	$(x+2)x(x-1)$
4	$\Psi_5(x)\Psi_8(x)$	$(x^2+x-1)(x^2-2)$
5	$[\Psi_2(x)]^2\Psi_3(x)\Psi_6(x)\Psi_{10}(x)$	$(x+2)(x+1)(x-1)(x^2-x-1)$
6	$\Psi_4(x)\Psi_7(x)\Psi_{12}(x)$	$x(x^3+x^2-2x-1)(x^2-3)$
7	$[\Psi_2(x)]^2\Psi_4(x)\Psi_8(x)\Psi_{14}(x)$	$(x+2)x(x^2-2)(x^3-x^2-2x+1)$
8	$\Psi_3(x)\Psi_9(x)\Psi_{16}(x)$	$(x+1)(x^3-3x+1)(x^4-4x^2+2)$
9	$[\Psi_2(x)]^2\Psi_5(x)\Psi_6(x)\Psi_{10}(x)\Psi_{18}(x)$	$(x+2)(x^2+x-1)(x-1)(x^2-x-1)(x^3-3x-1)$
10	$\Psi_4(x)\Psi_{11}(x)\Psi_{20}(x)$	$x(x^5+x^4-4x^3-3x^2+3x+1)(x^4-5x^2+5)$
11	$[\Psi_2(x)]^2\Psi_3(x)\Psi_4(x)\Psi_6(x)\Psi_{12}(x)$ $\Psi_{22}(x)$	$(x+2)(x+1)x(x-1)(x^2-3)$ $(x^5-x^4-4x^3+3x^2+3x-1)$
12	$\Psi_8(x)\Psi_{13}(x)\Psi_{24}(x)$	$(x^2-2)(x^6+x^5-5x^4-4x^3+6x^2+3x-1)(x^4-4x^2+1)$
13	$[\Psi_2(x)]^2\Psi_7(x)\Psi_{14}(x)\Psi_{26}(x)$	$(x+2)(x^3+x^2-2x-1)(x^3-x^2-2x+1)$ $(x^6-x^5-5x^4+4x^3+6x^2-3x-1)$
14	$\Psi_3(x)\Psi_4(x)\Psi_5(x)\Psi_{15}(x)\Psi_{28}(x)$	$(x+1)x(x^2+x-1)(x^4-x^3-4x^2+4x+1)$ $(x^6-7x^4+14x^2-7)$
15	$[\Psi_2(x)]^2\Psi_4(x)\Psi_6(x)\Psi_8(x)\Psi_{10}(x)$ $\Psi_{16}(x)\Psi_{30}(x)$	$(x+2)x(x-1)(x^2-2)(x^2-x-1)(x^4-4x^2+2)$ $(x^4+x^3-4x^2-4x+1)$
16	$\Psi_{17}(x)\Psi_{32}(x)$	$(x^8+x^7-7x^6-6x^5+15x^4+10x^3-10x^2-4x+1)$ $(x^8-8x^6+20x^4-16x^2+2)$
17	$[\Psi_2(x)]^2\Psi_3(x)\Psi_6(x)\Psi_9(x)\Psi_{18}(x)$ $\Psi_{34}(x)$	$(x+2)(x+1)(x-1)(x^3-3x+1)(x^3-3x-1)$ $(x^8-x^7-7x^6+6x^5+15x^4-10x^3-10x^2+4x+1)$
18	$\Psi_4(x)\Psi_{12}(x)\Psi_{19}(x)\Psi_{36}(x)$	$x(x^2-3)(x^9+x^8-8x^7-7x^6+21x^5+15x^4-20x^3-$ $-10x^2+5x+1)(x^6-6x^4+9x^2-3)$
19	$[\Psi_2(x)]^2\Psi_4(x)\Psi_5(x)\Psi_{10}(x)\Psi_{20}(x)$ $\Psi_{38}(x)$	$(x+2)x(x^2+x-1)(x^2-x-1)(x^4-5x^2+5)(x^9-x^8-$ $-8x^7+7x^6+21x^5-15x^4-20x^3+10x^2+5x-1)$
20	$\Psi_3(x)\Psi_7(x)\Psi_8(x)\Psi_{21}(x)\Psi_{40}(x)$	$(x+1)(x^3+x^2-2x-1)(x^2-2)(x^6-x^5-6x^4+6x^3+$ $+8x^2-8x+1)(x^8-8x^6+19x^4-12x^2+1)$

Table 29: Irreducible factorization of $T_n^{**}(x)$ for $1 \le n \le 20$

n	indirect	direct expression
1	$\Psi_4^{**}(x)$	x
2	$\Psi_8^{**}(x)$	x^2+2
3	$\Psi_4^{**}(x)\Psi_{12}^{**}(x)$	$x(x^2+3)$
4	$\Psi_{16}^{**}(x)$	x^4+4x^2+2
5	$\Psi_4^{**}(x)\Psi_{20}^{**}(x)$	$x(x^4+5x^2+5)$
6	$\Psi_8^{**}(x)\Psi_{24}^{**}(x)$	$(x^2+2)(x^4+4x^2+1)$
7	$\Psi_4^{**}(x)\Psi_{28}^{**}(x)$	$x(x^6+7x^4+14x^2+7)$
8	$\Psi_{32}^{**}(x)$	$x^8+8x^6+20x^4+16x^2+2$
9	$\Psi_4^{**}(x)\Psi_{12}^{**}(x)\Psi_{36}^{**}(x)$	$x(x^2+3)(x^6+6x^4+9x^2+3)$
10	$\Psi_8^{**}(x)\Psi_{40}^{**}(x)$	$(x^2+2)(x^8+8x^6+19x^4+12x^2+1)$
11	$\Psi_4^{**}(x)\Psi_{44}^{**}(x)$	$x(x^{10}+11x^8+44x^6+77x^4+55x^2+11)$
12	$\Psi_{16}^{**}(x)\Psi_{48}^{**}(x)$	$(x^4+4x^2+2)(x^8+8x^6+20x^4+16x^2+1)$
13	$\Psi_4^{**}(x)\Psi_{52}^{**}(x)$	$x(x^{12}+13x^{10}+65x^8+156x^6+182x^4+91x^2+13)$
14	$\Psi_8^{**}(x)\Psi_{56}^{**}(x)$	$(x^2+2)(x^{12}+12x^{10}+53x^8+104x^6+86x^4+24x^2+1)$
15	$\Psi_4^{**}(x)\Psi_{12}^{**}(x)\Psi_{20}^{**}(x)$ $\Psi_{60}^{**}(x)$	$x(x^2+3)(x^4+5x^2+5)(x^8+7x^6+14x^4+8x^2+1)$
16	$\Psi_{64}^{**}(x)$	$x^{16}+16x^{14}+104x^{12}+352x^{10}+660x^8+672x^6+336x^4+64x^2+2$
17	$\Psi_4^{**}(x)\Psi_{68}^{**}(x)$	$x(x^{16}+17x^{14}+119x^{12}+442x^{10}+935x^8+1122x^6+$ $+714x^4+204x^2+17)$
18	$\Psi_8^{**}(x)\Psi_{24}^{**}(x)\Psi_{72}^{**}(x)$	$(x^2+2)(x^4+4x^2+1)(x^{12}+12x^{10}+54x^8+112x^6+105x^4+36x^2+1)$
19	$\Psi_4^{**}(x)\Psi_{76}^{**}(x)$	$x(x^{18}+19x^{16}+152x^{14}+665x^{12}+1729x^{10}+2717x^8+$ $+2508x^6+1254x^4+285x^2+19)$
20	$\Psi_{16}^{**}(x)\Psi_{80}^{**}(x)$	$(x^4+4x^2+2)(x^{16}+16x^{14}+104x^{12}+352x^{10}+659x^8+664x^6+$ $+316x^4+48x^2+1)$

Table 30: Irreducible factorization of $T_n^{**}(x) - 2$ for $1 \leq n \leq 20$

n	indirect	direct expression
2	$[\Psi_4^{**}(x)]^2$	x^2
4	$[\Psi_1^{**}(x)]^2[\Psi_4^{**}(x)]^2$	$(x^2+4)x^2$
6	$[\Psi_4^{**}(x)]^2[\Psi_{12}^{**}(x)]^2$	$x^2(x^2+3)^2$
8	$[\Psi_1^{**}(x)]^2[\Psi_4^{**}(x)]^2[\Psi_8^{**}(x)]^2$	$(x^2+4)x^2(x^2+2)^2$
10	$[\Psi_4^{**}(x)]^2[\Psi_{20}^{**}(x)]^2$	$x^2(x^4+5x^2+5)^2$
12	$[\Psi_1^{**}(x)]^2[\Psi_3^{**}(x)]^2[\Psi_4^{**}(x)]^2[\Psi_{12}^{**}(x)]^2$	$(x^2+4)(x^2+1)^2x^2(x^2+3)^2$
14	$[\Psi_4^{**}(x)]^2[\Psi_{28}^{**}(x)]^2$	$x^2(x^6+7x^4+14x^2+7)^2$
16	$[\Psi_1^{**}(x)]^2[\Psi_4^{**}(x)]^2[\Psi_8^{**}(x)]^2\Psi_{16}^{**}(x)]^2$	$(x^2+4)x^2(x^2+2)^2(x^4+4x^2+2)^2$
18	$[\Psi_4^{**}(x)]^2[\Psi_{12}^{**}(x)]^2\Psi_{36}^{**}(x)]^2$	$x^2(x^2+3)^2(x^6+6x^4+9x^2+3)^2$
20	$[\Psi_1^{**}(x)]^2[\Psi_4^{**}(x)]^2[\Psi_5^{**}(x)]^2\Psi_{20}^{**}(x)]^2$	$(x^2+4)x^2(x^4+3x^2+1)^2(x^4+5x^2+5)^2$

Table 31: Irreducible factorization of $T_n^{**}(x) + 2$ for $1 \leq n \leq 20$

n	indirect	direct expression
2	$[\Psi_1^{**}(x)]^2$	x^2+4
4	$[\Psi_8^{**}(x)]^2$	$(x^2+2)^2$
6	$[\Psi_1^{**}(x)]^2[\Psi_3^{**}(x)]^2$	$(x^2+4)(x^2+1)^2$
8	$[\Psi_{16}^{**}(x)]^2$	$(x^4+4x^2+2)^2$
10	$[\Psi_1^{**}(x)]^2[\Psi_5^{**}(x)]^2$	$(x^2+4)(x^4+3x^2+1)^2$
12	$[\Psi_8^{**}(x)]^2[\Psi_{24}^{**}(x)]^2$	$(x^2+2)^2(x^4+4x^2+1)^2$
14	$[\Psi_1^{**}(x)]^2[\Psi_7^{**}(x)]^2$	$(x^2+4)(x^6+5x^4+6x^2+1)^2$
16	$[\Psi_{32}^{**}(x)]^2$	$(x^8+8x^6+20x^4+16x^2+2)^2$
18	$[\Psi_1^{**}(x)]^2[\Psi_3^{**}(x)]^2\Psi_9^{**}(x)]^2$	$(x^2+4)(x^2+1)^2(x^6+6x^4+9x^2+1)^2$
20	$[\Psi_8^{**}(x)]^2[\Psi_{40}^{**}(x)]^2$	$(x^2+2)^2(x^8+8x^6+19x^4+12x^2+1)^2$

Table 32: Irreducible factorization of $T_n^{**}(x)-1$ for $1 \le n \le 20$

n	indirect	direct expression
2	$\Psi_3^{**}(x)$	x^2+1
4	$\Psi_{24}^{**}(x)$	x^4+4x^2+1
6	$\Psi_9^{**}(x)$	$x^6+6x^4+9x^2+1$
8	$\Psi_{48}^{**}(x)$	$x^8+8x^6+20x^4+16x^2+1$
10	$\Psi_3^{**}(x)\Psi_{15}^{**}(x)$	$(x^2+1)(x^8+9x^6+26x^4+24x^2+1)$
12	$\Psi_{72}^{**}(x)$	$x^{12}+12x^{10}+54x^8+112x^6+105x^4+36x^2+1$
14	$\Psi_3^{**}(x)\Psi_{21}^{**}(x)$	$(x^2+1)(x^{12}+13x^{10}+64x^8+146x^6+148x^4+48x^2+1)$
16	$\Psi_{96}^{**}(x)$	$x^{16}+16x^{14}+104x^{12}+352x^{10}+660x^8+672x^6+336x^4+64x^2+1$
18	$\Psi_{27}^{**}(x)$	$x^{18}+18x^{16}+135x^{14}+546x^{12}+1287x^{10}+1782x^8+$ $+1386x^6+540x^4+81x^2+1$
20	$\Psi_{24}^{**}(x)\Psi_{120}^{**}(x)$	$(x^4+4x^2+1)(x^{16}+16x^{14}+105x^{12}+364x^{10}+714x^8+784x^6+$ $+440x^4+96x^2+1)$

Table 33: Irreducible factorization of $T_n^{**}(x)+1$ for $1 \le n \le 20$

n	indirect	direct expression
2	$\Psi_{12}^{**}(x)$	x^2+3
4	$\Psi_3^{**}(x)\Psi_{12}^{**}(x)$	$(x^2+1)(x^2+3)$
6	$\Psi_{36}^{**}(x)$	$x^6+6x^4+9x^2+3$
8	$\Psi_3^{**}(x)\Psi_{12}^{**}(x)\Psi_{24}^{**}(x)$	$(x^2+1)(x^2+3)(x^4+4x^2+1)$
10	$\Psi_{12}^{**}(x)\Psi_{60}^{**}(x)$	$(x^2+3)(x^8+7x^6+14x^4+8x^2+1)$
12	$\Psi_9^{**}(x)\Psi_{36}^{**}(x)$	$(x^6+6x^4+9x^2+1)(x^6+6x^4+9x^2+3)$
14	$\Psi_{12}^{**}(x)\Psi_{84}^{**}(x)$	$(x^2+3)(x^{12}+11x^{10}+44x^8+78x^6+60x^4+16x^2+1)$
16	$\Psi_3^{**}(x)\Psi_{12}^{**}(x)\Psi_{24}^{**}(x)\Psi_{48}^{**}(x)$	$(x^2+1)(x^2+3)(x^4+4x^2+1)(x^8+8x^6+20x^4+16x^2+1)$
18	$\Psi_{108}^{**}(x)$	$x^{18}+18x^{16}+135x^{14}+546x^{12}+1287x^{10}+1782x^8+$ $+1386x^6+540x^4+81x^2+3$
20	$\Psi_3^{**}(x)\Psi_{12}^{**}(x)\Psi_{15}^{**}(x)\Psi_{60}^{**}(x)$	$(x^2+1)(x^2+3)(x^8+9x^6+26x^4+24x^2+1)$ $(x^8+7x^6+14x^4+8x^2+1)$

Table 34: Irreducible factorization of $U_n^{**}(x)$ for $1 \le n \le 20$

n	indirect	direct expression
1	$\Psi_4^{**}(x)$	x
2	$\Psi_3^{**}(x)$	x^2+1
3	$\Psi_4^{**}(x)\Psi_8^{**}(x)$	$x(x^2+2)$
4	$\Psi_5^{**}(x)$	x^4+3x^2+1
5	$\Psi_3^{**}(x)\Psi_4^{**}(x)\Psi_{12}^{**}(x)$	$(x^2+1)x(x^2+3)$
6	$\Psi_7^{**}(x)$	$x^6+5x^4+6x^2+1$
7	$\Psi_4^{**}(x)\Psi_8^{**}(x)\Psi_{16}^{**}(x)$	$x(x^2+2)(x^4+4x^2+2)$
8	$\Psi_3^{**}(x)\Psi_9^{**}(x)$	$(x^2+1)(x^6+6x^4+9x^2+1)$
9	$\Psi_4^{**}(x)\Psi_5^{**}(x)\Psi_{20}^{**}(x)$	$x(x^4+3x^2+1)(x^4+5x^2+5)$
10	$\Psi_{11}^{**}(x)$	$x^{10}+9x^8+28x^6+35x^4+15x^2+1$
11	$\Psi_3^{**}(x)\Psi_4^{**}(x)\Psi_8^{**}(x)$ $\Psi_{12}^{**}(x)\Psi_{24}^{**}(x)$	$(x^2+1)x(x^2+2)(x^2+3)(x^4+4x^2+1)$
12	$\Psi_{13}^{**}(x)$	$x^{12}+11x^{10}+45x^8+84x^6+70x^4+21x^2+1$
13	$\Psi_4^{**}(x)\Psi_7^{**}(x)\Psi_{28}^{**}(x)$	$x(x^6+5x^4+6x^2+1)(x^6+7x^4+14x^2+7)$
14	$\Psi_3^{**}(x)\Psi_5^{**}(x)\Psi_{15}^{**}(x)$	$(x^2+1)(x^4+3x^2+1)(x^8+9x^6+26x^4+24x^2+1)$
15	$\Psi_4^{**}(x)\Psi_8^{**}(x)\Psi_{16}^{**}(x)\Psi_{32}^{**}(x)$	$x(x^2+2)(x^4+4x^2+2)(x^8+8x^6+20x^4+16x^2+2)$
16	$\Psi_{17}^{**}(x)$	$x^{16}+15x^{14}+91x^{12}+286x^{10}+495x^8+$ $+462x^6+210x^4+36x^2+1$
17	$\Psi_3^{**}(x)\Psi_4^{**}(x)\Psi_9^{**}(x)$ $\Psi_{12}^{**}(x)\Psi_{36}^{**}(x)$	$(x^2+1)x(x^6+6x^4+9x^2+1)(x^2+3)(x^6+6x^4+9x^2+3)$
18	$\Psi_{19}^{**}(x)$	$x^{18}+17x^{16}+120x^{14}+455x^{12}+1001x^{10}+1287x^8+$ $+924x^6+330x^4+45x^2+1$
19	$\Psi_4^{**}(x)\Psi_5^{**}(x)\Psi_8^{**}(x)$ $\Psi_{20}^{**}(x)\Psi_{40}^{**}(x)$	$x(x^4+3x^2+1)(x^2+2)(x^4+5x^2+5)$ $(x^8+8x^6+19x^4+12x^2+1)$
20	$\Psi_3^{**}(x)\Psi_7^{**}(x)\Psi_{21}^{**}(x)$	$(x^2+1)(x^6+5x^4+6x^2+1)$ $(x^{12}+13x^{10}+64x^8+146x^6+148x^4+48x^2+1)$

Table 35: Irreducible factorization of $U_n^{**}(x) - 1$ for $1 \le n \le 20$

n	indirect	direct expression
2	$[\Psi_4^{**}(x)]^2$	x^2
4	$[\Psi_4^{**}(x)]^2\Psi_{12}^{**}(x)$	$x^2(x^2+3)$
6	$[\Psi_4^{**}(x)]^2\Psi_8^{**}(x)\Psi_{12}^{**}(x)$	$x^2(x^2+2)(x^2+3)$
8	$[\Psi_4^{**}(x)]^2\Psi_8^{**}(x)\Psi_{20}^{**}(x)$	$x^2(x^2+2)(x^4+5x^2+5)$
10	$\Psi_3^{**}(x)[\Psi_4^{**}(x)]^2\Psi_{12}^{**}(x)\Psi_{20}^{**}(x)$	$(x^2+1)x^2(x^2+3)(x^4+5x^2+5)$
12	$\Psi_3^{**}(x)[\Psi_4^{**}(x)]^2\Psi_{12}^{**}(x)\Psi_{28}^{**}(x)$	$(x^2+1)x^2(x^2+3)(x^6+7x^4+14x^2+7)$
14	$[\Psi_4^{**}(x)]^2\Psi_8^{**}(x)\Psi_{16}^{**}(x)\Psi_{28}^{**}(x)$	$x^2(x^2+2)(x^4+4x^2+2)(x^6+7x^4+14x^2+7)$
16	$[\Psi_4^{**}(x)]^2\Psi_8^{**}(x)\Psi_{12}^{**}(x)\Psi_{16}^{**}(x)\Psi_{36}^{**}(x)$	$x^2(x^2+2)(x^2+3)(x^4+4x^2+2)(x^6+6x^4+9x^2+3)$
18	$[\Psi_4^{**}(x)]^2\Psi_5^{**}(x)\Psi_{12}^{**}(x)$ $\Psi_{20}^{**}(x)\Psi_{36}^{**}(x)$	$x^2(x^4+3x^2+1)(x^2+3)(x^4+5x^2+5)$ $(x^6+6x^4+9x^2+3)$
20	$[\Psi_4^{**}(x)]^2\Psi_5^{**}(x)\Psi_{20}^{**}(x)\Psi_{44}^{**}(x)$	$x^2(x^4+3x^2+1)(x^4+5x^2+5)$ $(x^{10}+11x^8+44x^6+77x^4+55x^2+11)$

Table 36: Irreducible factorization of $U_n^{**}(x) + 1$ for $1 \le n \le 20$

n	indirect	direct expression
2	$\Psi_8^{**}(x)$	x^2+2
4	$\Psi_3^{**}(x)\Psi_8^{**}(x)$	$(x^2+1)(x^2+2)$
6	$\Psi_3^{**}(x)\Psi_{16}^{**}(x)$	$(x^2+1)(x^4+4x^2+2)$
8	$\Psi_5^{**}(x)\Psi_{16}^{**}(x)$	$(x^4+3x^2+1)(x^4+4x^2+2)$
10	$\Psi_5^{**}(x)\Psi_8^{**}(x)\Psi_{24}^{**}(x)$	$(x^4+3x^2+1)(x^2+2)(x^4+4x^2+1)$
12	$\Psi_7^{**}(x)\Psi_8^{**}(x)\Psi_{24}^{**}(x)$	$(x^6+5x^4+6x^2+1)(x^2+2)(x^4+4x^2+1)$
14	$\Psi_7^{**}(x)\Psi_{32}^{**}(x)$	$(x^6+5x^4+6x^2+1)(x^8+8x^6+20x^4+16x^2+2)$
16	$\Psi_3^{**}(x)\Psi_9^{**}(x)\Psi_{32}^{**}(x)$	$(x^2+1)(x^6+6x^4+9x^2+1)(x^8+8x^6+20x^4+16x^2+2)$
18	$\Psi_3^{**}(x)\Psi_8^{**}(x)\Psi_9^{**}(x)\Psi_{40}^{**}(x)$	$(x^2+1)(x^2+2)(x^6+6x^4+9x^2+1)$ $(x^8+8x^6+19x^4+12x^2+1)$
20	$\Psi_8^{**}(x)\Psi_{11}^{**}(x)\Psi_{40}^{**}(x)$	$(x^2+2)(x^{10}+9x^8+28x^6+35x^4+15x^2+1)$ $(x^8+8x^6+19x^4+12x^2+1)$

Table 37: List of $\Psi_n(x)$ and $\Phi_n(x)$ for $1 \le n \le 120$

n	$\Psi_n(x)$ (cyclotomic pre-polynomial)	$\Phi_n(x)$ (cyclotomic polynomial)
1	$\sqrt{x-2}$	$x-1$
2	$\sqrt{x+2}$	$x+1$
3	$x+1$	$\sum_{k=0}^{2} x^k$
4	x	x^2+1
5	x^2+x-1	$\sum_{k=0}^{4} x^k$
6	$x-1$	$\sum_{k=0}^{2}(-1)^k x^k$
7	x^3+x^2-2x-1	$\sum_{k=0}^{6} x^k$
8	x^2-2	x^4+1
9	x^3-3x+1	$\sum_{k=0}^{2} x^{3k}$
10	x^2-x-1	$\sum_{k=0}^{4}(-1)^k x^k$
11	$x^5+x^4-4x^3-3x^2+3x+1$	$\sum_{k=0}^{10} x^k$
12	x^2-3	$\sum_{k=0}^{2}(-1)^k x^{2k}$
13	$x^6+x^5-5x^4-4x^3+6x^2+3x-1$	$\sum_{k=0}^{12} x^k$
14	x^3-x^2-2x+1	$\sum_{k=0}^{6}(-1)^k x^k$
15	$x^4-x^3-4x^2+4x+1$	$x^8-x^7+x^5-x^4+x^3-x+1$
16	x^4-4x^2+2	x^8+1
17	$x^8+x^7-7x^6-6x^5+15x^4+10x^3-10x^2-4x+1$	$\sum_{k=0}^{16} x^k$
18	x^3-3x-1	$\sum_{k=0}^{2}(-1)^k x^{3k}$
19	$x^9+x^8-8x^7-7x^6+21x^5+15x^4-20x^3-10x^2+5x+1$	$\sum_{k=0}^{18} x^k$
20	x^4-5x^2+5	$\sum_{k=0}^{4}(-1)^k x^{2k}$

n	$\Psi_n(x)$ (cyclotomic pre-polynomial) $\Phi_n(x)$ (cyclotomic polynomial)
21	$x^6 - x^5 - 6x^4 + 6x^3 + 8x^2 - 8x + 1$
	$x^{12} - x^{11} + x^9 - x^8 + x^6 - x^4 + x^3 - x + 1$
22	$x^5 - x^4 - 4x^3 + 3x^2 + 3x - 1$
	$\sum_{k=0}^{10}(-1)^k x^k$
23	$x^{11} + x^{10} - 10x^9 - 9x^8 + 36x^7 + 28x^6 - 56x^5 - 35x^4 + 35x^3 + 15x^2 - 6x - 1$
	$\sum_{k=0}^{22} x^k$
24	$x^4 - 4x^2 + 1$
	$\sum_{k=0}^{2}(-1)^k x^{4k}$
25	$x^{10} - 10x^8 + 35x^6 + x^5 - 50x^4 - 5x^3 + 25x^2 + 5x - 1$
	$\sum_{k=0}^{4} x^{5k}$
26	$x^6 - x^5 - 5x^4 + 4x^3 + 6x^2 - 3x - 1$
	$\sum_{k=0}^{12}(-1)^k x^k$
27	$x^9 - 9x^7 + 27x^5 - 30x^3 + 9x + 1$
	$\sum_{k=0}^{2} x^{9k}$
28	$x^6 - 7x^4 + 14x^2 - 7$
	$\sum_{k=0}^{6}(-1)^k x^{2k}$
29	$\sum_{k=0}^{6}(-1)^k \left[\binom{14-k}{k} x^{14-2k} + \binom{13-k}{k} x^{13-2k}\right] - 1$
	$\sum_{k=0}^{28} x^k$
30	$x^4 + x^3 - 4x^2 - 4x + 1$
	$x^8 + x^7 - x^5 - x^4 - x^3 + x + 1$
31	$\sum_{k=0}^{7}(-1)^k \left[\binom{15-k}{k} x^{15-2k} + \binom{14-k}{k} x^{14-2k}\right]$
	$\sum_{k=0}^{30} x^k$
32	$x^8 - 8x^6 + 20x^4 - 16x^2 + 2$
	$x^{16} + 1$

33	$x^{10} - x^9 - 10x^8 + 10x^7 + 34x^6 - 34x^5 - 43x^4 + 43x^3 + 12x^2 - 12x + 1$
	$x^{20} - x^{19} + x^{17} - x^{16} + x^{14} - x^{13} + x^{11} - x^{10} + x^9 - x^7 + x^6 - x^4 + x^3 - x + 1$
34	$x^8 - x^7 - 7x^6 + 6x^5 + 15x^4 - 10x^3 - 10x^2 + 4x + 1$
	$\sum_{k=0}^{16}(-1)^k x^k$
35	$x^{12} - x^{11} - 12x^{10} + 11x^9 + 54x^8 - 43x^7 - 113x^6 + 71x^5 + 110x^4 - 46x^3 - 40x^2 + 8x + 1$
	$x^{24} - x^{23} + x^{19} - x^{18} + x^{17} - x^{16} + x^{14} - x^{13} + x^{12} - x^{11} + x^{10} - x^8 + x^7 - x^6 + x^5 - x + 1$
36	$x^6 - 6x^4 + 9x^2 - 3$
	$\sum_{k=0}^{2}(-1)^k x^{6k}$
37	$\sum_{k=0}^{8}(-1)^k \left[\binom{18-k}{k}x^{18-2k} + \binom{17-k}{k}x^{17-2k}\right] - 1$
	$\sum_{k=0}^{36} x^k$
38	$x^9 - x^8 - 8x^7 + 7x^6 + 21x^5 - 15x^4 - 20x^3 + 10x^2 + 5x - 1$
	$\sum_{k=0}^{18}(-1)^k x^k$
39	$x^{12} - x^{11} - 12x^{10} + 12x^9 + 53x^8 - 53x^7 - 103x^6 + 103x^5 + 79x^4 - 79x^3 - 12x^2 + 12x + 1$
	$x^{24} - x^{23} + x^{21} - x^{20} + x^{18} - x^{17} + x^{15} - x^{14} + x^{12} - x^{10} + x^9 - x^7 + x^6 - x^4 +$ $+x^3 - x + 1$
40	$x^8 - 8x^6 + 19x^4 - 12x^2 + 1$
	$\sum_{k=0}^{4}(-1)^k x^{4k}$
41	$\sum_{k=0}^{9}(-1)^k \left[\binom{20-k}{k}x^{20-2k} + \binom{19-k}{k}x^{19-2k}\right] + 1$
	$\sum_{k=0}^{40} x^k$
42	$x^6 + x^5 - 6x^4 - 6x^3 + 8x^2 + 8x + 1$
	$x^{12} + x^{11} - x^9 - x^8 + x^6 - x^4 - x^3 + x + 1$
43	$\sum_{k=0}^{10}(-1)^k \left[\binom{21-k}{k}x^{21-2k} + \binom{20-k}{k}x^{20-2k}\right]$
	$\sum_{k=0}^{42} x^k$
44	$x^{10} - 11x^8 + 44x^6 - 77x^4 + 55x^2 - 11$
	$\sum_{k=0}^{10}(-1)^k x^{2k}$
45	$x^{12} - 12x^{10} - x^9 + 54x^8 + 9x^7 - 112x^6 - 27x^5 + 105x^4 + 31x^3 - 36x^2 - 12x + 1$
	$x^{24} - x^{21} + x^{15} - x^{12} + x^9 - x^3 + 1$

46	$x^{11}-x^{10}-10x^9+9x^8+36x^7-28x^6-56x^5+35x^4+35x^3-15x^2-6x+1$
	$\sum_{k=0}^{22}(-1)^k x^k$
47	$\sum_{k=0}^{11}(-1)^k\left[\binom{23-k}{k}x^{23-2k}+\binom{22-k}{k}x^{22-2k}\right]$
	$\sum_{k=0}^{46}x^k$
48	$x^8-8x^6+20x^4-16x^2+1$
	$\sum_{k=0}^{2}(-1)^k x^{8k}$
49	$x^{21}-21x^{19}+189x^{17}-952x^{15}+x^{14}+2940x^{13}-14x^{12}-5733x^{11}+77x^{10}+7007x^9-$ $-210x^8-5147x^7+294x^6+2072x^5-196x^4-371x^3+49x^2+14x-1$
	$\sum_{k=0}^{6}x^{7k}$
50	$x^{10}-10x^8+35x^6-x^5-50x^4+5x^3+25x^2-5x-1$
	$\sum_{k=0}^{4}(-1)^k x^{5k}$
51	$x^{16}-x^{15}-16x^{14}+16x^{13}+103x^{12}-103x^{11}-339x^{10}+339x^9+596x^8-596x^7-526x^6+$ $+526x^5+188x^4-188x^3-16x^2+16x+1$
	$x^{32}-x^{31}+x^{29}-x^{28}+x^{26}-x^{25}+x^{23}-x^{22}+x^{20}-x^{19}+x^{17}-x^{16}+x^{15}-x^{13}+x^{12}-$ $-x^{10}+x^9-x^7+x^6-x^4+x^3-x+1$
52	$x^{12}-13x^{10}+65x^8-156x^6+182x^4-91x^2+13$
	$\sum_{k=0}^{12}(-1)^k x^{2k}$
53	$\sum_{k=0}^{12}(-1)^k\left[\binom{26-k}{k}x^{26-2k}+\binom{25-k}{k}x^{25-2k}\right]-1$
	$\sum_{k=0}^{52}x^k$
54	$x^9-9x^7+27x^5-30x^3+9x-1$
	$\sum_{k=0}^{2}(-1)^k x^{9k}$
55	$x^{20}-x^{19}-20x^{18}+19x^{17}+170x^{16}-151x^{15}-801x^{14}+650x^{13}+2289x^{12}-1639x^{11}-$ $-4080x^{10}+2442x^9+4489x^8-2058x^7-2891x^6+877x^5+951x^4-151x^3-108x^2+12x+1$
	$x^{40}-x^{39}+x^{35}-x^{34}+x^{30}-x^{28}+x^{25}-x^{23}+x^{20}-x^{17}+x^{15}-x^{12}+x^{10}-x^6+x^5-$ $-x+1$
56	$x^{12}-12x^{10}+53x^8-104x^6+86x^4-24x^2+1$
	$\sum_{k=0}^{6}(-1)^k x^{4k}$

57	$x^{18}-x^{17}-18x^{16}+18x^{15}+134x^{14}-134x^{13}-531x^{12}+531x^{11}+1198x^{10}-1198x^9-$ $-1519x^8+1519x^7+989x^6-989x^5-265x^4+265x^3+20x^2-20x+1$ $x^{36}-x^{35}+x^{33}-x^{32}+x^{30}-x^{29}+x^{27}-x^{26}+x^{24}-x^{23}+x^{21}-x^{20}+x^{18}-x^{16}+$ $+x^{15}-x^{13}+x^{12}-x^{10}+x^9-x^7+x^6-x^4+x^3-x+1$
58	$\sum_{k=0}^{6}(-1)^k\left[\binom{14-k}{k}x^{14-2k}-\binom{13-k}{k}x^{13-2k}\right]-1$ $\sum_{k=0}^{28}(-1)^k x^k$
59	$\sum_{k=0}^{14}(-1)^k\left[\binom{29-k}{k}x^{29-2k}+\binom{28-k}{k}x^{28-2k}\right]$ $\sum_{k=0}^{58} x^k$
60	$x^8-7x^6+14x^4-8x^2+1$ $x^{16}+x^{14}-x^{10}-x^8-x^6+x^2+1$
61	$\sum_{k=0}^{14}(-1)^k\left[\binom{30-k}{k}x^{30-2k}+\binom{29-k}{k}x^{29-2k}\right]-1$ $\sum_{k=0}^{60} x^k$
62	$\sum_{k=0}^{7}(-1)^k\left[\binom{15-k}{k}x^{15-2k}-\binom{14-k}{k}x^{14-2k}\right]$ $\sum_{k=0}^{30}(-1)^k x^k$
63	$x^{18}-18x^{16}-x^{15}+135x^{14}+15x^{13}-546x^{12}-90x^{11}+1287x^{10}+276x^9-1782x^8-459x^7+$ $+1385x^6+405x^5-534x^4-170x^3+72x^2+24x+1$ $x^{36}-x^{33}+x^{27}-x^{24}+x^{18}-x^{12}+x^9-x^3+1$
64	$x^{16}-16x^{14}+104x^{12}-352x^{10}+660x^8-672x^6+336x^4-64x^2+2$ $x^{32}+1$
65	$x^{24}-x^{23}-24x^{22}+23x^{21}+252x^{20}-229x^{19}-1521x^{18}+1292x^{17}+5832x^{16}-4540x^{15}-$ $-14822x^{14}+10282x^{13}+25284x^{12}-15001x^{11}-28667x^{10}+13653x^9+20886x^8-7168x^7-$ $-9126x^6+1802x^5+2085x^4-101x^3-180x^2-12x+1$ $x^{48}-x^{47}+x^{43}-x^{42}+x^{38}-x^{37}+x^{35}-x^{34}+x^{33}-x^{32}+x^{30}-x^{29}+x^{28}-x^{27}+x^{25}-$ $-x^{24}+x^{23}-x^{21}+x^{20}-x^{19}+x^{18}-x^{16}+x^{15}-x^{14}+x^{13}-x^{11}+x^{10}-x^6+x^5-x+1$
66	$x^{10}+x^9-10x^8-10x^7+34x^6+34x^5-43x^4-43x^3+12x^2+12x+1$ $x^{20}+x^{19}-x^{17}-x^{16}+x^{14}+x^{13}-x^{11}-x^{10}-x^9+x^7+x^6-x^4-x^3+x+1$
67	$\sum_{k=0}^{16}(-1)^k\left[\binom{33-k}{k}x^{33-2k}+\binom{32-k}{k}x^{32-2k}\right]$ $\sum_{k=0}^{66} x^k$

68	$x^{16} - 17x^{14} + 119x^{12} - 442x^{10} + 935x^8 - 1122x^6 + 714x^4 - 204x^2 + 17$
	$\sum_{k=0}^{16}(-1)^k x^{2k}$
69	$x^{22} - x^{21} - 22x^{20} + 22x^{19} + 208x^{18} - 208x^{17} - 1103x^{16} + 1103x^{15} + 3589x^{14} - 3589x^{13} -$ $-7359x^{12} + 7359x^{11} + 9385x^{10} - 9385x^9 - 7060x^8 + 7060x^7 + 2807x^6 - 2807x^5 - 482x^4 +$ $+482x^3 + 24x^2 - 24x + 1$
	$x^{44} - x^{43} + x^{41} - x^{40} + x^{38} - x^{37} + x^{35} - x^{34} + x^{32} - x^{31} + x^{29} - x^{28} + x^{26} - x^{25} + x^{23} -$ $-x^{22} + x^{21} - x^{19} + x^{18} - x^{16} + x^{15} - x^{13} + x^{12} - x^{10} + x^9 - x^7 + x^6 - x^4 + x^3 - x + 1$
70	$x^{12} + x^{11} - 12x^{10} - 11x^9 + 54x^8 + 43x^7 - 113x^6 - 71x^5 + 110x^4 + 46x^3 - 40x^2 - 8x + 1$
	$x^{24} + x^{23} - x^{19} - x^{18} - x^{17} - x^{16} + x^{14} + x^{13} + x^{12} + x^{11} + x^{10} - x^8 - x^7 - x^6 -$ $-x^5 + x + 1$
71	$\sum_{k=0}^{17}(-1)^k \left[\binom{35-k}{k}x^{35-2k} + \binom{34-k}{k}x^{34-2k}\right]$
	$\sum_{k=0}^{70} x^k$
72	$x^{12} - 12x^{10} + 54x^8 - 112x^6 + 105x^4 - 36x^2 + 1$
	$\sum_{k=0}^{2}(-1)^k x^{12k}$
73	$\sum_{k=0}^{17}(-1)^k \left[\binom{36-k}{k}x^{36-2k} + \binom{35-k}{k}x^{35-2k}\right] + 1$
	$\sum_{k=0}^{72} x^k$
74	$\sum_{k=0}^{8}(-1)^k \left[\binom{18-k}{k}x^{18-2k} - \binom{17-k}{k}x^{17-2k}\right] - 1$
	$\sum_{k=0}^{36}(-1)^k x^k$
75	$x^{20} - 20x^{18} + 170x^{16} - x^{15} - 800x^{14} + 15x^{13} + 2275x^{12} - 90x^{11} - 4004x^{10} + 275x^9 + 4290x^8 -$ $-450x^7 - 2640x^6 + 379x^5 + 825x^4 - 145x^3 - 100x^2 + 20x + 1$
	$x^{40} - x^{35} + x^{25} - x^{20} + x^{15} - x^5 + 1$
76	$x^{18} - 19x^{16} + 152x^{14} - 665x^{12} + 1729x^{10} - 2717x^8 + 2508x^6 - 1254x^4 + 285x^2 - 19$
	$\sum_{k=0}^{18}(-1)^k x^{2k}$
77	$x^{30} - x^{29} - 30x^{28} + 29x^{27} + 405x^{26} - 377x^{25} - 3250x^{24} + 2901x^{23} + 17249x^{22} - 14697x^{21} -$ $-63734x^{20} + 51590x^{19} + 168035x^{18} - 128611x^{17} - 318629x^{16} + 229651x^{15} + 432159x^{14} -$ $-292608x^{13} - 411241x^{12} + 261924x^{11} + 264472x^{10} - 159873x^9 - 107406x^8 + 63143x^7 + 24051x^6 -$ $-14609x^5 - 2010x^4 + +1562x^3 - 72x^2 - 24x + 1$
	$x^{60} - x^{59} + x^{53} - x^{52} + x^{49} - x^{48} + x^{46} - x^{45} + x^{42} - x^{41} + x^{39} - x^{37} + x^{35} - x^{34} +$ $+x^{32} - x^{30} + x^{28} - x^{26} + x^{25} - x^{23} + x^{21} - x^{19} + x^{18} - x^{15} + x^{14} - x^{12} + x^{11} - x^8 +$ $+x^7 - x + 1$

78	$x^{12}+x^{11}-12x^{10}-12x^9+53x^8+53x^7-103x^6-103x^5+79x^4+79x^3-12x^2-12x+1$
	$x^{24}+x^{23}-x^{21}-x^{20}+x^{18}+x^{17}-x^{15}-x^{14}+x^{12}-x^{10}-x^9+x^7+x^6-x^4-$ $-x^3+x+1$
79	$\sum_{k=0}^{19}(-1)^k\left[\binom{39-k}{k}x^{39-2k}+\binom{38-k}{k}x^{38-2k}\right]$
	$\sum_{k=0}^{78}x^k$
80	$x^{16}-16x^{14}+104x^{12}-352x^{10}+659x^8-664x^6+316x^4-48x^2+1$
	$\sum_{k=0}^{4}(-1)^k x^{8k}$
81	$x^{27}-27x^{25}+324x^{23}-2277x^{21}+10395x^{19}-32319x^{17}+69768x^{15}-104652x^{13}+107406x^{11}-$ $-72930x^9+30888x^7-7371x^5+819x^3-27x+1$
	$\sum_{k=0}^{2}x^{27k}$
82	$\sum_{k=0}^{9}(-1)^k\left[\binom{20-k}{k}x^{20-2k}-\binom{19-k}{k}x^{19-2k}\right]+1$
	$\sum_{k=0}^{40}(-1)^k x^k$
83	$\sum_{k=0}^{20}(-1)^k\left[\binom{41-k}{k}x^{41-2k}+\binom{40-k}{k}x^{40-2k}\right]$
	$\sum_{k=0}^{82}x^k$
84	$x^{12}-11x^{10}+44x^8-78x^6+60x^4-16x^2+1$
	$x^{24}+x^{22}-x^{18}-x^{16}+x^{12}-x^8-x^6+x^2+1$
85	$x^{32}-x^{31}-32x^{30}+31x^{29}+464x^{28}-433x^{27}-4033x^{26}+3600x^{25}+23426x^{24}-19826x^{23}-$ $-95978x^{22}+76152x^{21}+285340x^{20}-209188x^{19}-623732x^{18}+414544x^{17}+1004699x^{16}-$ $-590154x^{15}-1183593x^{14}+593422x^{13}+1001431x^{12}-407890x^{11}-589403x^{10}+181071x^9+$ $+228888x^8-46882x^7-53690x^6+5686x^5+6516x^4-116x^3-304x^2-16x+1$
	$x^{64}-x^{63}+x^{59}-x^{58}+x^{54}-x^{53}+x^{49}-x^{48}+x^{47}-x^{46}+x^{44}-x^{43}+x^{42}-x^{41}+x^{39}-$ $-x^{38}+x^{37}-x^{36}+x^{34}-x^{33}+x^{32}-x^{31}+x^{30}-x^{28}+x^{27}-x^{26}+x^{25}-x^{23}+x^{22}-$ $-x^{21}+x^{20}-x^{18}+x^{17}-x^{16}+x^{15}-x^{11}+x^{10}-x^6+x^5-x+1$
86	$\sum_{k=0}^{10}(-1)^k\left[\binom{21-k}{k}x^{21-2k}-\binom{20-k}{k}x^{20-2k}\right]$
	$\sum_{k=0}^{42}(-1)^k x^k$
87	$x^{28}-x^{27}-28x^{26}+28x^{25}+349x^{24}-349x^{23}-2551x^{22}+2551x^{21}+12123x^{20}-12123x^{19}-$ $-39236x^{18}+39236x^{17}+88045x^{16}-88045x^{15}-136763x^{14}+136763x^{13}+144247x^{12}-$ $-144247x^{11}-99295x^{10}+99295x^9+41703x^8-41703x^7-9569x^6+9569x^5+987x^4-$ $-987x^3-28x^2+28x+1$
	$x^{56}-x^{55}+x^{53}-x^{52}+x^{50}-x^{49}+x^{47}-x^{46}+x^{44}-x^{43}+x^{41}-x^{40}+x^{38}-x^{37}+x^{35}-$ $-x^{34}+x^{32}-x^{31}+x^{29}-x^{28}+x^{27}-x^{25}+x^{24}-x^{22}+x^{21}-x^{19}+x^{18}-x^{16}+x^{15}-$ $-x^{13}+x^{12}-x^{10}+x^9-x^7+x^6-x^4+x^3-x+1$

88 $x^{20} - 20x^{18} + 169x^{16} - 784x^{14} + 2172x^{12} - 3664x^{10} + 3683x^{8} - 2072x^{6} + 575x^{4} - 60x^{2} + 1$

$\sum_{k=0}^{10}(-1)^k x^{4k}$

89 $\sum_{k=0}^{21}(-1)^k \left[\binom{44-k}{k}x^{44-2k} + \binom{43-k}{k}x^{43-2k}\right] + 1$

$\sum_{k=0}^{88} x^k$

90 $x^{12} - 12x^{10} + x^{9} + 54x^{8} - 9x^{7} - 112x^{6} + 27x^{5} + 105x^{4} - 31x^{3} - 36x^{2} + 12x + 1$

$x^{24} + x^{21} - x^{15} - x^{12} - x^{9} + x^{3} + 1$

91 $x^{36} - x^{35} - 36x^{34} + 35x^{33} + 594x^{32} - 560x^{31} - 5952x^{30} + 5426x^{29} + 40454x^{28} - 35554x^{27} - 197288x^{26} + 166634x^{25} + 712180x^{24} - 576199x^{23} - 1934944x^{22} + 1494700x^{21} + 3983738x^{20} - 2928983x^{19} - 6208136x^{18} + 4331906x^{17} + 7259749x^{16} - 4795428x^{15} - 6263633x^{14} + 3907332x^{13} + 3879994x^{12} - 2278575x^{11} - 1655004x^{10} + 908896x^{9} + 456052x^{8} - 230058x^{7} - 73571x^{6} + 32327x^{5} + 6018x^{4} - 1922x^{3} - 216x^{2} + 24x + 1$

$x^{72} - x^{71} + x^{65} - x^{64} + x^{59} - x^{57} + x^{52} - x^{50} + x^{46} - x^{43} + x^{39} - x^{36} + x^{33} - x^{29} + x^{26} - x^{22} + x^{20} - x^{15} + x^{13} - x^{8} + x^{7} - x + 1$

92 $x^{22} - 23x^{20} + 230x^{18} - 1311x^{16} + 4692x^{14} - 10948x^{12} + 16744x^{10} - 16445x^{8} + 9867x^{6} - 3289x^{4} + 506x^{2} - 23$

$\sum_{k=0}^{22}(-1)^k x^{2k}$

93 $x^{30} - x^{29} - 30x^{28} + 30x^{27} + 404x^{26} - 404x^{25} - 3223x^{24} + 3223x^{23} + 16927x^{22} - 16927x^{21} - 61503x^{20} + 61503x^{19} + 158101x^{18} - 158101x^{17} - 288950x^{16} + 288950x^{15} + 371908x^{14} - 371908x^{13} - 329002x^{12} + 329002x^{11} + 191674x^{10} - 191674x^{9} - 68664x^{8} + 68664x^{7} + 13548x^{6} - 13548x^{5} - 1208x^{4} + 1208x^{3} + 32x^{2} - 32x + 1$

$x^{60} - x^{59} + x^{57} - x^{56} + x^{54} - x^{53} + x^{51} - x^{50} + x^{48} - x^{47} + x^{45} - x^{44} + x^{42} - x^{41} + x^{39} - x^{38} + x^{36} - x^{35} + x^{33} - x^{32} + x^{30} - x^{28} + x^{27} - x^{25} + x^{24} - x^{22} + x^{21} - x^{19} + x^{18} - x^{16} + x^{15} - x^{13} + x^{12} - x^{10} + x^{9} - x^{7} + x^{6} - x^{4} + x^{3} - x + 1$

94 $\sum_{k=0}^{11}(-1)^k \left[\binom{23-k}{k}x^{23-2k} - \binom{22-k}{k}x^{22-2k}\right]$

$\sum_{k=0}^{46}(-1)^k x^k$

95 $x^{36} - x^{35} - 36x^{34} + 35x^{33} + 594x^{32} - 559x^{31} - 5953x^{30} + 5394x^{29} + 40485x^{28} - 35091x^{27} - 197720x^{26} + 162629x^{25} + 715754x^{24} - 553125x^{23} - 1954471x^{22} + 1401346x^{21} + 4057888x^{20} - 2656542x^{19} - 6408680x^{18} + 3752139x^{17} + 7649116x^{16} - 3896996x^{15} - 6803518x^{14} + 2906674x^{13} + 4404908x^{12} - 1498899x^{11} - 2000649x^{10} + 503479x^{9} + 601194x^{8} - 100432x^{7} - 108458x^{6} + 10534x^{5} + 9885x^{4} - 605x^{3} - 340x^{2} + 20x + 1$

$x^{72} - x^{71} + x^{67} - x^{66} + x^{62} - x^{61} + x^{57} - x^{56} + x^{53} - x^{51} + x^{48} - x^{46} + x^{43} - x^{41} + x^{38} - x^{36} + x^{34} - x^{31} + x^{29} - x^{26} + x^{24} - x^{21} + x^{19} - x^{16} + x^{15} - x^{11} + x^{10} - x^{6} + x^{5} - x + 1$

96 $x^{16} - 16x^{14} + 104x^{12} - 352x^{10} + 660x^{8} - 672x^{6} + 336x^{4} - 64x^{2} + 1$

$\sum_{k=0}^{2}(-1)^k x^{16k}$

97	$\sum_{k=0}^{23}(-1)^k\left[\binom{48-k}{k}x^{48-2k}+\binom{47-k}{k}x^{47-2k}\right]+1$ $\sum_{k=0}^{96}x^k$
98	$x^{21}-21x^{19}+189x^{17}-952x^{15}-x^{14}+2940x^{13}+14x^{12}-5733x^{11}-77x^{10}+7007x^9+210x^8-$ $-5147x^7-294x^6+2072x^5+196x^4-371x^3-49x^2+14x+1$ $\sum_{k=0}^{6}(-1)^kx^{7k}$
99	$x^{30}-30x^{28}-x^{27}+405x^{26}+27x^{25}-3250x^{24}-324x^{23}+17250x^{22}+2278x^{21}-63756x^{20}-$ $-10416x^{19}+168244x^{18}+32508x^{17}-319752x^{16}-70720x^{15}+435915x^{14}+107592x^{13}-$ $-419353x^{12}-113139x^{11}+275835x^{10}+79936x^9-117504x^8-36027x^7+29442x^6+$ $+9423x^5-3555x^4-1173x^3+108x^2+36x+1$ $x^{60}-x^{57}+x^{51}-x^{48}+x^{42}-x^{39}+x^{33}-x^{30}+x^{27}-x^{21}+x^{18}-x^{12}+x^9-x^3+1$
100	$x^{20}-20x^{18}+170x^{16}-800x^{14}+2275x^{12}-4005x^{10}+4300x^8-2675x^6+875x^4-125x^2+5$ $\sum_{k=0}^{4}(-1)^kx^{10k}$
101	$\sum_{k=0}^{24}(-1)^k\left[\binom{50-k}{k}x^{50-2k}+\binom{49-k}{k}x^{49-2k}\right]-1$ $\sum_{k=0}^{100}x^k$
102	$x^{16}+x^{15}-16x^{14}-16x^{13}+103x^{12}+103x^{11}-339x^{10}-339x^9+596x^8+596x^7-526x^6-$ $-526x^5+188x^4+188x^3-16x^2-16x+1$ $x^{32}+x^{31}-x^{29}-x^{28}+x^{26}+x^{25}-x^{23}-x^{22}+x^{20}+x^{19}-x^{17}-x^{16}-x^{15}+x^{13}+x^{12}-$ $-x^{10}-x^9+x^7+x^6-x^4-x^3+x+1$
103	$\sum_{k=0}^{25}(-1)^k\left[\binom{51-k}{k}x^{51-2k}+\binom{50-k}{k}x^{50-2k}\right]$ $\sum_{k=0}^{102}x^k$
104	$x^{24}-24x^{22}+251x^{20}-1500x^{18}+5645x^{16}-13904x^{14}+22580x^{12}-23792x^{10}+15622x^8-$ $-5936x^6+1141x^4-84x^2+1$ $\sum_{k=0}^{12}(-1)^kx^{4k}$
105	$x^{24}+x^{23}-23x^{22}-23x^{21}+230x^{20}+229x^{19}-1312x^{18}-1294x^{17}+4709x^{16}+4573x^{15}-$ $-11067x^{14}-10506x^{13}+17187x^{12}+15810x^{11}-17391x^{10}-15333x^9+11034x^8+9189x^7-$ $-4088x^6-3142x^5+784x^4+528x^3-64x^2-32x+1$ $x^{48}+x^{47}+x^{46}-x^{43}-x^{42}-2x^{41}-x^{40}-x^{39}+x^{36}+x^{35}+x^{34}+x^{33}+x^{32}+x^{31}-$ $-x^{28}-x^{26}-x^{24}-x^{22}-x^{20}+x^{17}+x^{16}+x^{15}+x^{14}+x^{13}+x^{12}-x^9-x^8-2x^7-$ $-x^6-x^5+x^2+x+1$
106	$\sum_{k=0}^{12}(-1)^k\left[\binom{26-k}{k}x^{26-2k}-\binom{25-k}{k}x^{25-2k}\right]-1$ $\sum_{k=0}^{52}(-1)^kx^k$

107	$\sum_{k=0}^{26}(-1)^k\left[\binom{53-k}{k}x^{53-2k}+\binom{52-k}{k}x^{52-2k}\right]$ $\sum_{k=0}^{106}x^k$
108	$x^{18}-18x^{16}+135x^{14}-546x^{12}+1287x^{10}-1782x^8+1386x^6-540x^4+81x^2-3$ $x^{36}-x^{18}+1$
109	$\sum_{k=0}^{26}(-1)^k\left[\binom{54-k}{k}x^{54-2k}+\binom{53-k}{k}x^{53-2k}\right]-1$ $\sum_{k=0}^{108}x^k$
110	$x^{20}+x^{19}-20x^{18}-19x^{17}+170x^{16}+151x^{15}-801x^{14}-650x^{13}+2289x^{12}+1639x^{11}-$ $-4080x^{10}-2442x^9+4489x^8+2058x^7-2891x^6-877x^5+951x^4+151x^3-108x^2-12x+1$ $x^{40}+x^{39}-x^{35}-x^{34}+x^{30}-x^{28}-x^{25}+x^{23}+x^{20}+x^{17}-x^{15}-x^{12}+x^{10}-x^6-$ $-x^5+x+1$
111	$x^{36}-x^{35}-36x^{34}+36x^{33}+593x^{32}-593x^{31}-5919x^{30}+5919x^{29}+39961x^{28}-39961x^{27}-$ $-192880x^{26}+192880x^{25}+685907x^{24}-685907x^{23}-1824913x^{22}+1824913x^{21}+$ $+3651272x^{20}-3651272x^{19}-5475703x^{18}+5475703x^{17}+6085132x^{16}-6085132x^{15}-$ $-4909788x^{14}+4909788x^{13}+2786656x^{12}-2786656x^{11}-1061566x^{10}+1061566x^9+$ $+253044x^8-253044x^7-33780x^6+33780x^5+2073x^4-2073x^3-36x^2+36x+1$ $x^{72}-x^{71}+x^{69}-x^{68}+x^{66}-x^{65}+x^{63}-x^{62}+x^{60}-x^{59}+x^{57}-x^{56}+x^{54}-x^{53}+x^{51}-$ $-x^{50}+x^{48}-x^{47}+x^{45}-x^{44}+x^{42}-x^{41}+x^{39}-x^{38}+x^{36}-x^{34}+x^{33}-x^{31}+x^{30}-$ $-x^{28}+x^{27}-x^{25}+x^{24}-x^{22}+x^{21}-x^{19}+x^{18}-x^{16}+x^{15}-x^{13}+x^{12}-x^{10}+x^9-$ $-x^7+x^6-x^4+x^3-x+1$
112	$x^{24}-24x^{22}+252x^{20}-1520x^{18}+5813x^{16}-14672x^{14}+24648x^{12}-27104x^{10}+18646x^8-$ $-7344x^6+1400x^4-96x^2+1$ $\sum_{k=0}^{6}(-1)^k x^{8k}$
113	$\sum_{k=0}^{27}(-1)^k\left[\binom{56-k}{k}x^{56-2k}+\binom{55-k}{k}x^{55-2k}\right]+1$ $\sum_{k=0}^{112}x^k$
114	$x^{18}+x^{17}-18x^{16}-18x^{15}+134x^{14}+134x^{13}-531x^{12}-531x^{11}+1198x^{10}+1198x^9-$ $-1519x^8-1519x^7+989x^6+989x^5-265x^4-265x^3+20x^2+20x+1$ $x^{36}+x^{35}-x^{33}-x^{32}+x^{30}+x^{29}-x^{27}-x^{26}+x^{24}+x^{23}-x^{21}-x^{20}+x^{18}-x^{16}-x^{15}+$ $+x^{13}+x^{12}-x^{10}-x^9+x^7+x^6-x^4-x^3+x+1$

115	$x^{44} - x^{43} - 44x^{42} + 43x^{41} + 902x^{40} - 859x^{39} - 11441x^{38} + 10582x^{37} + 100567x^{36} - 89985x^{35} - 650236x^{34} + 560251x^{33} + 3203650x^{32} - 2643399x^{31} - 12294569x^{30} + 9651170x^{29} + 37253225x^{28} - 27602055x^{27} - 89808680x^{26} + 62206625x^{25} + 172779011x^{24} - 110572386x^{23} - 265002643x^{22} + 154430258x^{21} + 322457859x^{20} - 168027624x^{19} - 308473797x^{18} + 140446403x^{17} + 228768475x^{16} - 88323383x^{15} - 128871012x^{14} + 40552321x^{13} + 53558102x^{12} - 13016729x^{11} - 15742859x^{10} + 2742874x^{9} + 3073662x^{8} - 347233x^{7} - 361455x^{6} + 24089x^{5} + 21786x^{4} - 986x^{3} - 504x^{2} + 24x + 1$
	$x^{88} - x^{87} + x^{83} - x^{82} + x^{78} - x^{77} + x^{73} - x^{72} + x^{68} - x^{67} + x^{65} - x^{64} + x^{63} - x^{62} + x^{60} - x^{59} + x^{58} - x^{57} + x^{55} - x^{54} + x^{53} - x^{52} + x^{50} - x^{49} + x^{48} - x^{47} + x^{45} - x^{44} + x^{43} - x^{41} + x^{40} - x^{39} + x^{38} - x^{36} + x^{35} - x^{34} + x^{33} - x^{31} + x^{30} - x^{29} + x^{28} - x^{26} + x^{25} - x^{24} + x^{23} - x^{21} + x^{20} - x^{16} + x^{15} - x^{11} + x^{10} - x^{6} + x^{5} - x + 1$
116	$x^{28} - 29x^{26} + 377x^{24} - 2900x^{22} + 14674x^{20} - 51359x^{18} + 127281x^{16} - 224808x^{14} + 281010x^{12} - 243542x^{10} + 140998x^{8} - 51272x^{6} + 10556x^{4} - 1015x^{2} + 29$
	$\sum_{k=0}^{28}(-1)^k x^{2k}$
117	$x^{36} - 36x^{34} - x^{33} + 594x^{32} + 33x^{31} - 5952x^{30} - 495x^{29} + 40455x^{28} + 4467x^{27} - 197316x^{26} - 27054x^{25} + 712529x^{24} + 116154x^{23} - 1937496x^{22} - 364067x^{21} + 3995883x^{20} + 845295x^{19} - 6247579x^{18} - 1459998x^{17} + 7348878x^{16} + 1867585x^{15} - 6403833x^{14} - 1746123x^{13} + 4030936x^{12} + 1165464x^{11} - 1761969x^{10} - 534544x^{9} + 502173x^{8} + 158337x^{7} - 83631x^{6} - 27162x^{5} + 6471x^{4} + 2145x^{3} - 108x^{2} - 36x + 1$
	$x^{72} - x^{69} + x^{63} - x^{60} + x^{54} - x^{51} + x^{45} - x^{42} + x^{36} - x^{30} + x^{27} - x^{21} + x^{18} - x^{12} + x^{9} - x^{3} + 1$
118	$\sum_{k=0}^{14}(-1)^k \left[\binom{29-k}{k}x^{29-2k} - \binom{28-k}{k}x^{28-2k}\right]$
	$\sum_{k=0}^{58}(-1)^k x^k$
119	$x^{48} - x^{47} - 48x^{46} + 47x^{45} + 1080x^{44} - 1034x^{43} - 15136x^{42} + 14148x^{41} + 148091x^{40} - 134931x^{39} - 1074488x^{38} + 952717x^{37} + 5994444x^{36} - 5163498x^{35} - 26311919x^{34} + 21979367x^{33} + 92220138x^{32} - 74573322x^{31} - 260447090x^{30} + 203520550x^{29} + 595550236x^{28} - 448955699x^{27} - 1103970115x^{26} + 801604023x^{25} + 1655619772x^{24} - 1156350782x^{23} - 1998238943x^{22} + 1341017921x^{21} + 1924029811x^{20} - 1239775442x^{19} - 1459008975x^{18} + 902369642x^{17} + 855733436x^{16} - 507960152x^{15} - 378612663x^{14} + 215647349x^{13} + 122048196x^{12} - 66587715x^{11} - 27295100x^{10} + 14169142x^{9} + 3952116x^{8} - 1909570x^{7} - 338276x^{6} + 141792x^{5} + 15784x^{4} - 4584x^{3} - 352x^{2} + 32x + 1$
	$x^{96} - x^{95} + x^{89} - x^{88} + x^{82} - x^{81} + x^{79} - x^{78} + x^{75} - x^{74} + x^{72} - x^{71} + x^{68} - x^{67} + x^{65} - x^{64} + x^{62} - x^{60} + x^{58} - x^{57} + x^{55} - x^{53} + x^{51} - x^{50} + x^{48} - x^{46} + x^{45} - x^{43} + x^{41} - x^{39} + x^{38} - x^{36} + x^{34} - x^{32} + x^{31} - x^{29} + x^{28} - x^{25} + x^{24} - x^{22} + x^{21} - x^{18} + x^{17} - x^{15} + x^{14} - x^{8} + x^{7} - x + 1$
120	$x^{16} - 16x^{14} + 105x^{12} - 364x^{10} + 714x^{8} - 784x^{6} + 440x^{4} - 96x^{2} + 1$
	$x^{32} + x^{28} - x^{20} - x^{16} - x^{12} + x^{4} + 1$

Table 38: List of $\Psi_n^{**}(x) = |\Psi_n(ix)\Psi_{2n}(ix)| = |\Psi_n(ix)\Psi_n(-ix)|$ for $1 \leq n \leq 27,\ n$ odd

n	$\Psi_n^{**}(x)$
1	$\sqrt{x^2+4}$
3	x^2+1
5	x^4+3x^2+1
7	$x^6+5x^4+6x^2+1$
9	$x^6+6x^4+9x^2+1$
11	$x^{10}+9x^8+28x^6+35x^4+15x^2+1$
13	$x^{12}+11x^{10}+45x^8+84x^6+70x^4+21x^2+1$
15	$x^8+9x^6+26x^4+24x^2+1$
17	$x^{16}+15x^{14}+91x^{12}+286x^{10}+495x^8+462x^6+210x^4+36x^2+1$
19	$x^{18}+17x^{16}+120x^{14}+455x^{12}+1001x^{10}+1287x^8+924x^6+330x^4+45x^2+1$
21	$x^{12}+13x^{10}+64x^8+146x^6+148x^4+48x^2+1$
23	$x^{22}+21x^{20}+190x^{18}+969x^{16}+3060x^{14}+6188x^{12}+8008x^{10}+6435x^8+$ $+3003x^6+715x^4+66x^2+1$
25	$x^{20}+20x^{18}+170x^{16}+800x^{14}+2275x^{12}+4003x^{10}+4280x^8+2605x^6+775x^4+75x^2+1$
27	$x^{18}+18x^{16}+135x^{14}+546x^{12}+1287x^{10}+1782x^8+1386x^6+540x^4+81x^2+1$

Figure 1: Roots of $T_n^*(x) - 2$ and $\overline{T_n^*(x) - 2} = (x^n - 1)^2$

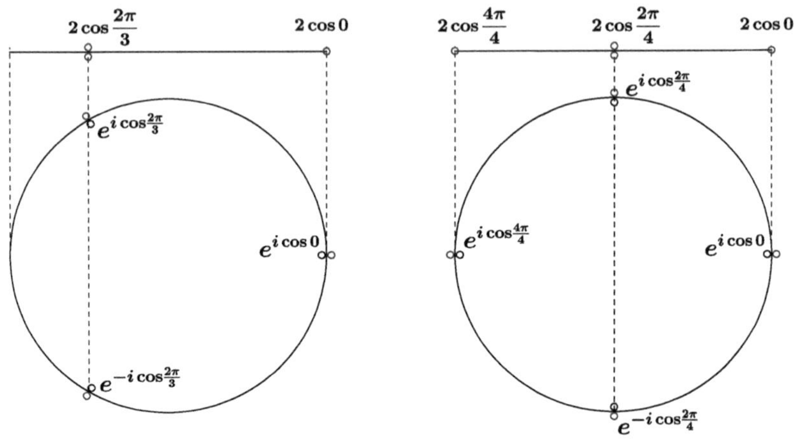

Figure 2: Roots of $T_n^*(x) + 2$ and $\overline{T_n^*(x) + 2} = (x^n + 1)^2$

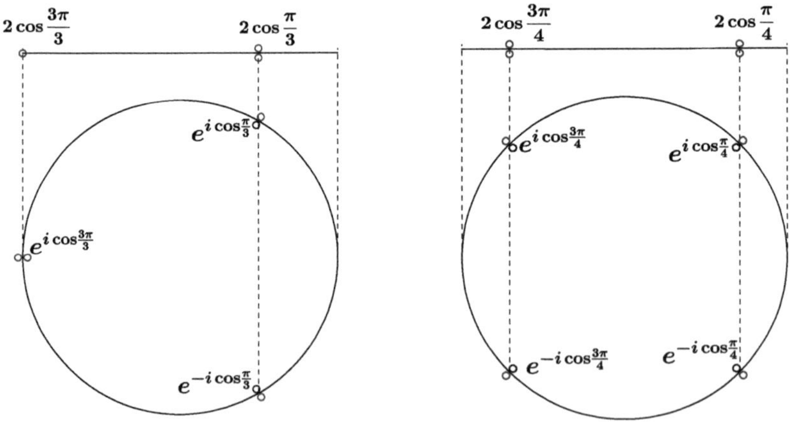

Figure 3: Roots of $T_n^*(x)$ and $\overline{T_n^*(x)} = x^{2n} + 1$

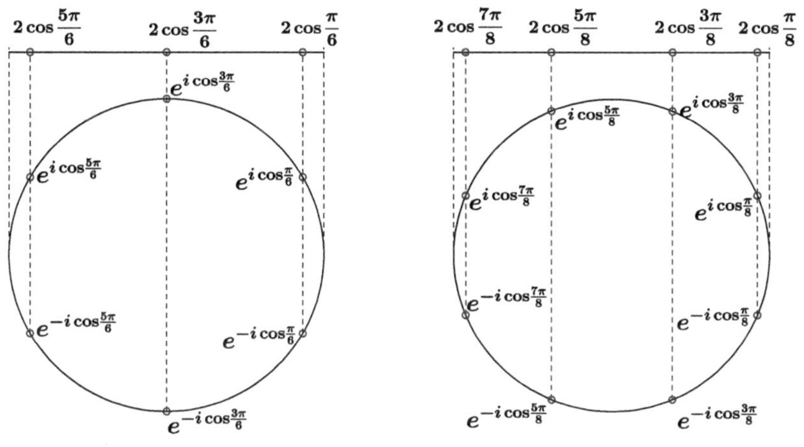

Figure 4: Roots of $T_n^*(x) - 1$ and $\overline{T_n^*(x) - 1} = x^{2n} - x^n + 1$

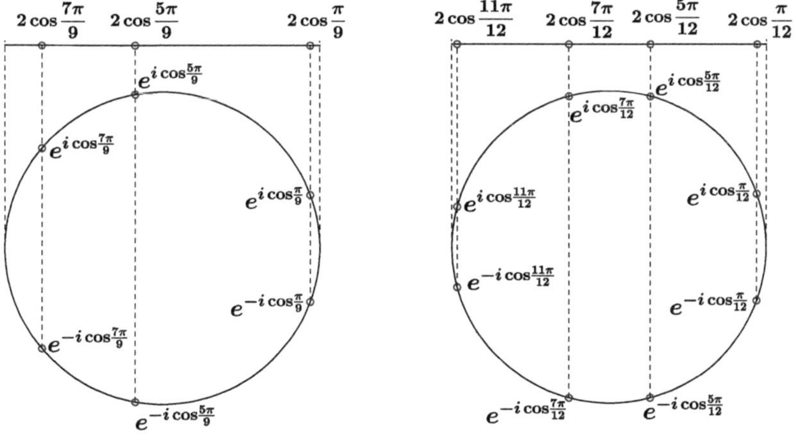

Figure 5: Roots of $T_n^*(x) + 1$ and $\overline{T_n^*(x) + 1} = x^{2n} + x^n + 1$

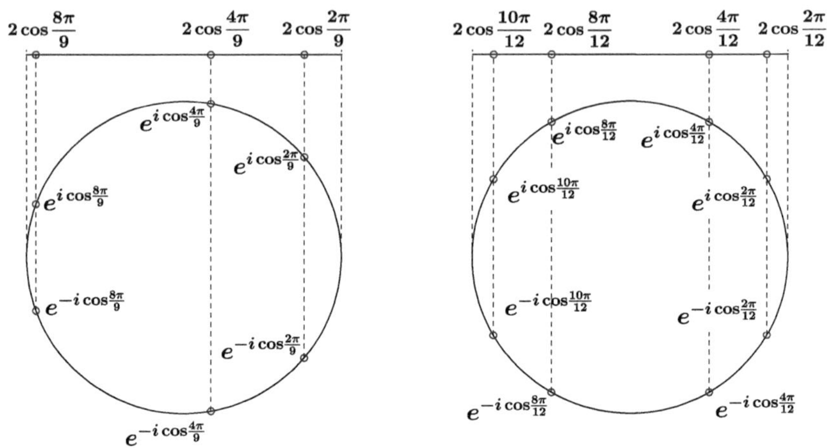

Figure 6: Roots of $U_n^*(x)$ and $\overline{U_n^*(x)} = \sum_{k=0}^{n} x^{2k}$

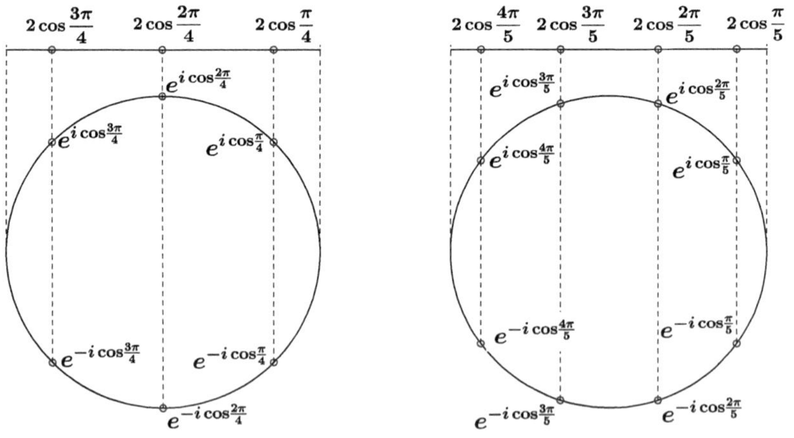

Figure 7: Roots of $U_n^*(x) - 1$ and $\overline{U_n^*(x) - 1} = \sum_{k=0}^{n} x^{2k} - x^n$

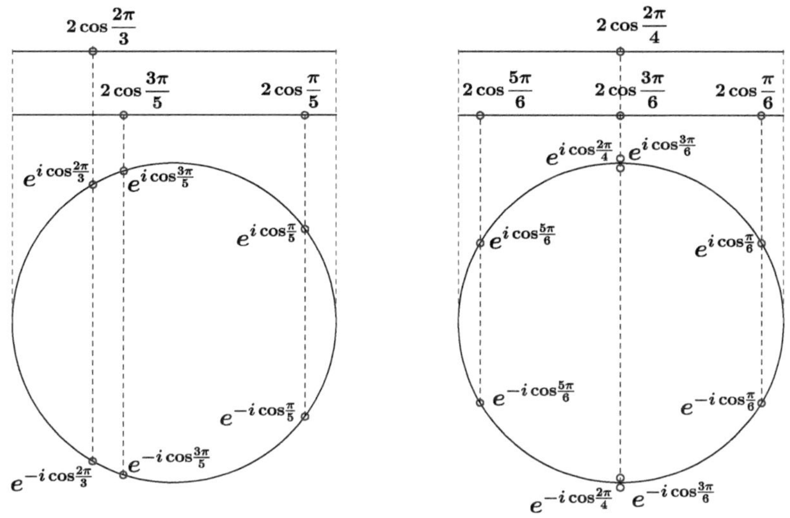

Figure 8: Roots of $U_n^*(x) + 1$ and $\overline{U_n^*(x) + 1} = \sum_{k=0}^{n} x^{2k} + x^n$

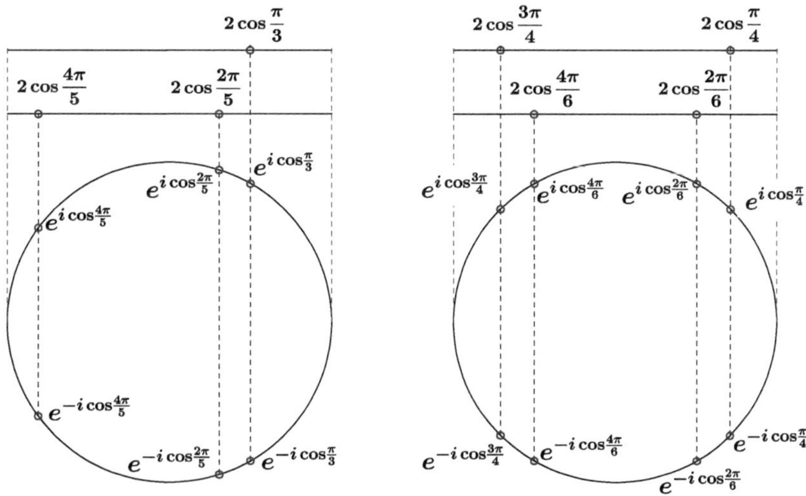

Bibliography

[1] Belbachir, H. and Szalay, L. *Fibonacci, and Lucas Pascal triangles*, (2016). Fibonacci, and Lucas Pascal triangles, Hacettepe Journal of Mathematics and Statistics **45** (5) (2016) 1231–1240.

[2] Kearnes, K. Kiss, E. and Szendrei, Á. *Gauss-egészek és Dirichlet tétele, 2. rész*, Középiskolai Matematikai és Fizikai Lapok (2010 április).
https://ewkiss.web.elte.hu/wp/wordpress/wp-content/uploads/2014/11/Dirichlet2.pdf

[3] Kéri, G. *Többszörös szögek táblázatai*, MTA SZTAKI, 1997.
http://old.sztaki.hu/~keri/math/mult.htm

[4] Kéri, G. *Compressed Chebyshev Polynomials and Multiple-Angle Formulas*, Omniscriptum Publishing Group, 2021.

[5] Mason, J. C. and Handscomb, D. C. *Chebyshev Polynomials*, Taylor and Francis, 2002.

[6] Niven, I. M. *Irrational Numbers*, The Carus Mathematical Monographs, No. 11. The Mathematical Association of America, Wiley, New York (1956).

[7] Wituła, R. and Słota, D. *On modified Chebyshev polynomials*, J. Math. Anal. Appl. **324** (2006) 321–343.

[8] Wikipedia, the free encyclopedia: Abel–Ruffini theorem
https://en.wikipedia.org/wiki/Abel–Ruffini_theorem

[9] Wikipedia, the free encyclopedia: Cyclotomic polynomial
https://en.wikipedia.org/wiki/Cyclotomic_polynomial

[10] Wikipedia, the free encyclopedia: Euler's totient function
https://en.wikipedia.org/wiki/Euler's_totient_function

[11] Wolfram MathWorld: Lucas Polynomial
https://mathworld.wolfram.com/LucasPolynomial.html

[12] eMathHelp: Equation Solver
https://www.emathhelp.net/calculators/algebra-2/equation-solver-calculator/

[13] Factoring Polynomials Calculator - eMathHelp
https://www.emathhelp.net/calculators/algebra-1/factoring-polynomials-calculator/

[14] Polynomial Factorization
https://www.dcode.fr/polynomial-factorization

Printed by Books on Demand GmbH, Norderstedt / Germany